Dr. John Sk

Key Terms
in Technology Studies

Wood, Metal and Plastic Technologies, Electricity/Electronics, Technical Graphics and Photography

Elbrook Press
P O Box 520
Brighton Road
South Australia 5048.

Elbrook Press
c/o Smith's Export Services
Carr Hill Industrial Estate
Dyform Works (Gate 9)
Balby
Doncaster DN4 8DG
England.

Elbrook Press
P O Box 51428
Pakuranga
Auckland
New Zealand

First published 1990

Copyright © John Skull, 1990

Designed by John Skull

ISBN 0 9587501 14

Typeset by The Government Printing Division, South Australia

Printed by Singapore National Printers Ltd., Singapore

Book cover by David Malpas and Martin Rowe

Illustrated by Alistair Best, Peter Hollard, David Malpas, Lisa Jarvis, Margaret Rees and Sara Rivers

No part of this publication may be reproduced, stored in a retrieval system, or transmitted in any form or by any means electronic, mechanical, photocopying recording or otherwise, without the prior approval of the publisher.

Acknowledgments

We are most grateful to the following people for their advice and help in the production of this book: Terry Smith in particular and also Terry Carr, Peter Collins, Peter Goldsmith, Keith Goldsworthy, Brian Haines, Bill Jolley, Derek Lambert, Fred Littlejohn, Robert Pilcher, Donald Schumacher and Stephen Wallace.

We are grateful also to the following organisations which have provided us with visual material: General Electric-Plastics; IBM Quarterly; Key Industries, South Australia; the Welding Institute, Abington, U.K.

If, inadvertently, we have failed to acknowledge any copyright material, we shall be happy to do so on receiving instructions.

Foreword

Technology in this book means:

the application of mainly scientific knowledge and expertise (often involving the use of machines) but also other forms of organised knowledge (such as expert skills and crafts and an understanding of social and cultural factors of a community) to practical tasks, the outcome of which will improve the lives of people.

The rationale of this book is based on the following opinions:

The beginning of understanding and wisdom begins with the precise meaning of terms.

> **(Confucius)**

As he uses words a person notices or neglects types of relationships and phenomena; he channels his reasoning and builds the house of consciousness.

> **(B.L.Whorf in** *Language, Thought and Reality*)

It is a basic tenet of education that what teachers do not teach students are not likely to learn.

What we cannot speak about, we consign to silence.

> **(L. Wittgenstein in** *Tractus Logico-Philosophicus*)

For if words are not things, they are living powers by which things of most importance to mankind are activated, combined and humanified.

> **(Samuel Taylor Coleridge in** *Preface to Aids to Reflection*)

We not only speak but think and even dream in words. Language is a mirror in which the whole spiritual development of mankind reflects itself. Therefore, in tracing words to their origins, we are tracing simultaneously civilization and culture to their real roots.

> **(Dr. Ernest Klein in** *A Comprehensive Etymological Dictionary of the English Language*)

It seems to be one of the paradoxes of creativity that in order to think originally, we must familiarise ourselves with the ideas of others.

> **(George Keller in** *The Art and Science of Creativity*)

He who has imagination without learning has wings but no feet.

> **(Chinese proverb)**

Abrasive

Pronounced: A-BRA-SIV (*1st a as in ago, 2nd a as in say, i as in ink*)

Origin

Abrasive is from the Latin *abrasus* which is from the verb *abdradere* meaning *to scrape off*. **To abrade** is to scrape off by vigorous rubbing.
Silica, flint and quartz (all of which mean the same thing) in the form of sand of varying textures have been used as an abrasive to smooth wood, metal and stone for more than 4000 years. The rough skins of shark and other fish were also used. The Chinese used crushed sea shells. Sandpaper, consisting of crushed glass glued to a coarse paper was first made in Switzerland in 1400. The process of making **carborundum**, a man-made abrasive, was evolved by Edward Goodrich Acheson of the U.S.A. in 1891. The first sanding machine was used in about 1877.

Meaning

An abrasive is a very hard substance used for rubbing or grinding a softer material, in order to shape the material, remove its surface or to make its surface smooth. An effective abrasive should be harder than the material being ground, be heat resistant so that it does not lose its effectiveness at grinding temperatures, be strong enough to withstand grinding pressures and be friable so that when a cutting edge becomes dull it will break to provide a new cutting edge.

sharpening a gouge on an abrasive wheel

Abrasive substances can be natural (e.g. corundum, diamond, garnet and flint) or man-made (e.g. manufactured diamond, aluminium oxide, silicon carbide, boron carbide and cubic boron nitride). **Corundum** is crystallised alumina (aluminium oxide), which is a mineral second only to diamond in hardness. The tradename **carborundum** was coined from the combination of carbon and corundum, and is also known as **silicon carbide**. **Emery** is a corundum.
The term abrasive refers not only to **sandpaper** (also called **garnet paper**), **emery paper**, **glass paper**, **steel wool** and **pumice stone** (made from volcanic rock) but also to tools, such as a **file**, a **rasp**, an **emery wheel**, a **grindstone**, an **oilstone** and various kinds of **power sanders** (e.g. drum sanders, belt sanders, orbital sanders, disc sanders, spindle sanders and portable sanders).

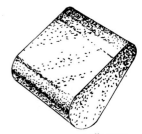
a slip stone

5

sanding on a lathe

Most **grinders** have abrasive wheels made from aluminium oxide or silicon carbide.

Associations

A **crocus cloth** is made from crushed iron oxide adhered to a strong fabric base. It is used for polishing metal. **Almandite** is a variety of garnet which is not as abrasive as carborundum or alumina. It is used for wood finishes. **Rouge**, a red oxide of iron, is a very fine abrasive which is used for polishing metal. **Tripoli** is a fine abrasive made from powdered silica. It is used for buffing metals. **Whiting** is a fine abrasive used for polishing metals. It is made from crushed chalk. **Abrasive blasting**, which can be a wet or dry process, is used for cleaning or finishing materials. Abrasive powder is blown by compressed air against an object.

Lapping is a process where very fine abrasives, such as diamond dust, are used to produce smooth finishes. A **buffing compound** is an abrasive which is made relatively gentle by being bound in wax. It is used for polishing metal. **Tooth** is used to describe the rough or absorbent quality of a surface which helps adhesion. **Coated abrasives** are papers or cloths coated with crushed and graded **grits**, such as glass, aluminium oxide, silicon carbide and garnet. These abrasives are available in a range of textures from fine to extra coarse. **Glass paper**, for instance, ranges from grade 00 which is very fine ,through 0, 1, 1.5 and F2 in the fine category, M2 and S2 in the medium category, 2.5, in the coarse category up to 3 which is extra coarse. The term *sanding* is used when using coated abrasives, although sand is not now used as an abrasive. The sanding of top coats of finishes is sometimes called **cutting back** or **flatting**, where there is a light scuff back to remove particles of dust.

See: **buffing, finish, glaze, grind, honing, lap, pinning, shape, whetting.**

using an orbital sander

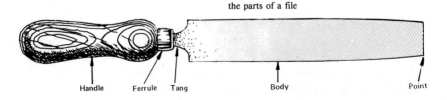

the parts of a file

Handle   Ferrule   Tang   Body   Point

Accelerator

Pronounced: AK-SEL-A-RA-TA (*1st a as in cat, 2nd a as in ago, 3rd a as in late, 4th a as in ago, e as in bell*)

Origin

From the Latin *acceleratio* meaning *to increase speed.* The Latin word *celer* means *swift* and *celerity* in English means *swiftness.*

Meaning

An accelerator is a substance which when added to other substances speeds up the chemical reaction. If salt, for example, is added to plaster, it accelerates the setting of the plaster. Most synthetic resins used as an adhesive need an acid hardener, which accelerates the speed of the adhesive's setting time. Heat is also used as an accelerator, such as in the curing of resins in the processing of veneers, where the heat is generated by high-frequency electrical waves.

An accelerator (e.g. an alkali, such as sodium carbonate) added to a photographic developer increases the activity of a developing agent. **Catalysts** lower the energy needed for a chemical reaction and hence accelerate the rate at which a reaction takes place. Consequently, they can be described as **accelerators.**

Associations

A **retarder** or **inhibitor** is a substance which decreases the speed of chemical reactions and can slow down the setting process. Keraton, for example, when added to plaster retards the setting time. An accelerator is sometimes called an **activator** or **promotor.** A **siccative** is a substance which is added to oil paint or to varnish to accelerate drying time.

See: **adhesive, buffing, catalyst, curing, develop, enamel, hygroscopic, lacquer, paint, varnish, veneer**

Adhesive

Pronounced: A-DEE-SIV (*a as in ago, ee as in see, i as in ink*)

Origin

From the French *adhesif* or *adhésive* (the feminine of

the word), which came from the Latin *adhaerere* meaning *to stick* or *cling*. One of the earliest forms of the many adhesives which now exist was **glue**, which derives from the Old French *glu* and the Latin *gluten* , which meant *birdlime* and *sticky paste*. Glue made from the boiled bones, horns, hoofs and hides of animals was used by the carpenters and cabinet makers in Ancient Egypt more than 5000 years ago. The use of synthetic adhesives made from chemicals has developed enormously during this century.

Meaning

applying adhesive

Adhesives are substances made either from things in nature, such as animals or plants, or from synthetic, man-made resins derived from chemicals. Their main function is to make the surface of materials (particularly fibres or particles) stick together and to remain firmly in contact. Occasionally, they are used to strengthen joints in joinery or carpentry and also to act as a **filler**.

Today there is a wide range of adhesives to choose from and it is important, to get the best results, that one chooses an adhesive which is best suited to a material under specific conditions for a particular function. A few of the many adhesives currently in use are: casein glue, vegetable glue, p.v.a. glue (polyvinyl acetate emulsions), urea formaldehyde glue, melamine formaldehyde glue, phenol formaldehyde, resorcinol formaldehyde glue, epoxy resin glue and contact (*"impact"*) glue. Most synthetic resins require the addition of a precise amount of acid *hardener*, which accelerates their setting time. This chemical reaction is called **polycondensation**. **Thermosetting adhesives** (e.g. epoxy and alkyds, which are often used to bond metals) harden by a non-reversible chemical reaction; **thermoplastic adhesives** (often used to bond metal to non-metals, such as wood, leather, paper, and cork) are hardened by cooling and softened by heating; **elastomeric adhesives** are made of latex or natural rubber and the addition of sulphur to them and the application of heat results in their **vulcanization** which increases their bonding strength. The process of hardening adhesives is called **curing**.

epoxy resin adhesives

Mucilage is a gum solution obtained from plant seeds.
Bleeding through describes the penetration of an adhesive through the face of a veneer which has been adhered

to a board material, which results in the discoloration of the veneer.

Bonding is used as an alternative term to adhesion.

Associations

See: **cramps, curing, filler, hardboard, hardener, hydraulic, key, laminate, resin, synthetic, thermoplastic, thermosetting, veneer, vulcanise.**

Alloy

Pronounced: A-LOY (*a as in cat,oy as in boy*)

Origin

From the Old French *aloi* which came from the verb *aloier* meaning *to bind, tie, fasten or combine.*

Meaning

An alloy is a metallic substance made by the fusing or melting of a **base metal** and at least one other metal or non-metal to form a new metal, which has different characteristics from the original metals. A **base metal** is known as a *pure* metal but an alloy is an *impure* metal, in that it comprises more than a base metal.

Different alloys have different functions. For instance, **babbit**, consisting of tin, copper and antimony, has special anti-frictional properties, so it is often used for lining moveable engine parts; **invar** (consisting of nickel, iron and a small amounts of manganese, silicon and carbon) has an almost zero rate of thermal expansion at normal temperatures, which makes it a very useful metal in the making of pendulums and balances in clocks and pistons in cars. **Bismuth** is used as an alloy, as it has a low melting-point and can be used in fire-security systems, as a substantial rise in temperature will melt the bismuth which can trigger the turning on of a sprinkler system. **Beryllium bronze** is a non-ferrous alloy of mainly copper with about 2% of beryllium. It is used for making non-sparking safety wrenches which can be safely used with oxy-acetylene welding equipment. **Tobin bronze** is a trade name for an alloy of copper, zinc, tin, iron and lead, which is used for many fittings on ships because of its high resistance to corrosion. Steels which have more than 4% of chromium in them are called **stainless steel**, because of their ability to resist stain and corrosion and weathering. **Speculum metal** consists of two parts copper and one part tin. As it has a bright, high-polish finish, it is used in the making of reflectors.

Alloys used in jewellery-making and casting are: white metal (about nine parts tin to one part antimony), pinchbeck metal (mainly copper and some zinc), German silver (mainly copper with varying amounts of nickel and zinc), casting brass (about six parts copper to four parts zinc and a little tin), pewter (nine parts tin with some antimony (7%) and a little copper) and electrum (equal parts of gold and silver). Other common alloys are: brass (copper and zinc), steel (iron and carbon), bronze (copper and tin), pewter (mainly tin, with some antimony and copper), and gar-alloy (zinc, copper and silver), which is used as a pewter substitute, as it is less expensive.

Associations

Amalgam is an alloy of mercury with one or more metals.
See: **alloy, brazing, corrosion, hardening, metallurgy, noble metals, solder, wrench**

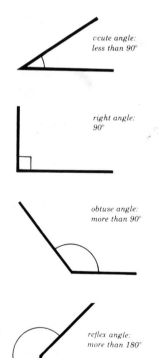

acute angle: less than 90°

right angle: 90°

obtuse angle: more than 90°

reflex angle: more than 180°

Angle

Pronounced: AN-GUL (*a as in cat, u as in bonus*)

Origin

From the Latin *angulus* meaning a *corner* or *bend*.

Meaning

An angle is the form produced by the coming together, or to the intersection of, two lines to an end point. The space between such lines which makes the figure is the angle. The four basic angles are: (a) a right angle (90°), (b) an acute angle (between 0 - 90°), (c) an obtuse angle (more than 90° but less than 180°) and (d) a reflex angle (more than 180°). An angle is **adjacent** to another angle if one leg is common to both angles. Two angles are **complementary** when their sum is 90°. Two angles are **supplementary** when their sum is 180°.

To place or fasten anything at an angle is to **skew** it.

Associations

A **scalene triangle** has no two sides equal; an **isosceles** triangle has two sides equal; an **equilateral triangle** has all sides equal; the **hypotenuse** is the side opposite to a right angle of a triangle. A **bisector** is a line which divides an angle (or object) into two equal parts.

A **gusset** or **angle bracket** is used to brace or stiffen a corner of a piece of work.

An angle plate is an L shaped piece of cast iron or steel machined at an angle of 90°. It is used to hold work at right angles on a marking-out or machine table.

See: **arris, bevel, chamfer, circuit, cross section, drawing, facet, fillet, foreshortening, mechanical drawing, mitre, parallax, pitch, polygon, polyhedron, projection, rake.**

Triangles

right   equilateral

isosceles   scalene

Annealing

Pronounced: AN-EE-LING (*a as in pan, ee as in bee, i as in sing*)

Origin

From an old English word *al* meaning *a fire*, from which the verb *onaelan* came, meaning *to burn, bake* or *set on fire.*

Meaning

Annealing is a process of reducing the stresses caused by heating metal (or glass) by a process of controlled heating and cooling. A metal is heated to a point above its usual hardening temperature and it is held at that temperature for sufficient time for the heat to spread evenly. It is then cooled slowly in a fire or in a sealed, temperature-controlled *muffle furnace.*

Metal which is softened by heat and then slowly cooled is less brittle and more easily worked (e.g. hammered, twisted or bent). Most metals become hardened when beaten or worked and need to be softened by an annealing process to enable further shaping. This process toughens and **tempers** or **glazes** the materials. With silver and copper, cooling must be done quickly by **quenching** them in cold water. Each metal requires a different and specific temperature for annealing and a specific period for which the temperature is retained. The process varies, depending upon whether the metal is ferrous or non-ferrous.

Associations

Blue annealing is a process of annealing where a bluish-black finish is given to sheets of steel. **Bright annealing** (or **box annealing**) is a process of annealing where *oxidisation* and discoloration are avoided by using containers to promote *reduction* (That is where a metal is starved of oxygen so that it is forced to give up its own oxygen to allow combustion).
See: **cladding, finish, flux, glaze, oxidation, pickle, temper.**

Anodise

Pronounced: ANO-DIZ (*a as in bat, o as in ago, i as in rise*)

Origin

From the Greek *anodos* meaning *the way up*. Anode was first used as a term in electricity by the chemist Michael Faraday (1791-1867).

Kind permission of Key Industries, Adelaide.

testing for anodic thickness using a permascope.

Meaning

To anodise (which is also called **anodic oxidation** or **anodic treatment**) has two main functions, namely, protective and/or decorative. For example, aluminium (which is a soft metal) can be covered with a hard, protective, durable film (e.g. with aluminium oxide of a thickness ranging from 0.005 to 0.02 mm.) which will oxidise it and prevent corrosion on such things as aeroplanes, ships and the exterior of buildings. The aluminium also can be made attractive, as the film (which is very porous) can be dyed any colour or tone (using coal-tar based or vegetable-based dyes) and the finish can be gloss or satin. Anodising is carried out in an acid solution of chromic or sulphuric acid electrolytically, using a process similar to that in electrolysis. Either AC or DC power can be used. The **anodic** film produced is usually sealed (to retain colouring and prevent **leaching**, its *washing away*) by boiling or steaming the metal in a steam chest for about 15 minutes at a temperature of 100°C. to form a hydrated aluminium oxide on it. The oxide film can be polished.

Associations:

An **anode** is the electrode through which an electric

current enters an electrolytic cell. It is the positive pole of the cell, and is usually the means through which an oxide is deposited during a plating process. The opposite of an anode is a **cathode**, which is a negative electrode.

See: **corrosion, electrode, electrolysis, finish, lipping, negative, oxidation, porosity, positive.**

Arbor

Pronounced: A-BAW (*a as in far, aw as in saw*)

Origin

From the French *arbre* meaning *tree* or *axis*, which came from the Latin *arbor* meaning *a tree*. In former days, a tree was usually used as an **axle** (the spindle upon which a wheel revolves and the rod connecting two wheels).

Meaning

An arbor is an **axle**, **shaft** or **spindle** on a machine which revolves and transmits mechanical force to other parts of the machine. It is the spindle which holds a cutting device on a milling machine and the shaft on which work is held for turning on a lathe. It is also called a **mandrel**.

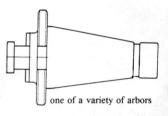
one of a variety of arbors

Associations

See: **lathe, mill, turn.**

Arris

Pronounced: ARIS (*a as in cat, i as in pin*)

Origin

From the Latin *arista* meaning *an ear of grain* (which has sharp edges) and the Old French *areste* meaning *a sharp edge or ridge*. The word *arête* now means a sharp crest of a mountain.

Meaning

An arris is the point at which two plane surfaces meet to form a sharp edge, such as the edge of a table top. If the sharp edge could be dangerous, as on a chair leg, table edge or shelf edge, the edge is shaved or sanded off. This process is called *removing the arris edge*.

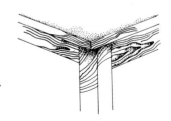

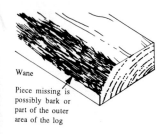

Wane
Piece missing is possibly bark or part of the outer area of the log

Associations

A **want edge** or **wane edge** is a piece of wood which is missing from an arris or an edge, owing to a sawing fault, the splitting of the wood or because bark or a piece of wood has been sheared off accidentally.
See: **angle, bevel**

Bauhaus

Pronounced: BOW-HOWS (*both ow's as in cow*)

Origin

The German word *bau* means *construction* and *haus* means *house*. The Bauhaus was a place (or house) of construction and creativity.

Meaning

In 1907 a professional organisation was established in Germany called the **German Werkbund** (meaning *the coming together of people* (bund) *to work*), whose aim was to produce simple, functional, very well-designed buildings, furniture and furnishing *using machines*. The individuality of a single skilled person using only his or her hands was superseded by a group of people working together using machine techniques.

Bauhaus refers to a School of Design and Architecture and Craft founded by a German, Walter Gropius, in Wermar, Germany, in 1919 and which closed in Berlin in 1933. The Bauhaus was much influenced by the Werkbund School of Arts and Crafts. It aimed to unite the arts and to combine the ideals of skill in craft and good design with mass production by machines and to close the gap between the creative artist and the industrial craftsperson. The functionalist approach of the Bauhaus is summed up in the statement "*form follows function*".

the Bauhaus logo

Associations

Associated with the Bauhaus School were famous artists, such as Wassily Kandinsky (Russian 1866-1944), Paul Klee (Swiss 1879-1940) and Lyonel Feininger (American 1871-1956).

Peter Behrens (1868 - 1940), one of the founders of the Werkbund, is usually acknowledged as the first professional industrial designer. Lazlo Moholy-Nagy (Hun-

garian 1895-1946) has probably had the greatest influence on the development of basic design principles in this century.
See: **design, furniture, Industrial Design**.

Bead

Pronounced: BEED (*ee as in seed*)

Origin

From the Old English *bed* meaning a *prayer*. In the saying of prayers in some religions, each prayer is counted by moving a small round perforated object along a string. The circular objects became known as *beads*. Beads then came to describe any round objects with a hole in their centres which were usually threaded on a string or wire to form a necklace. Then, any design or pattern on a moulding which used a succession of spherical or semi-spherical objects in a line became known as *beading*

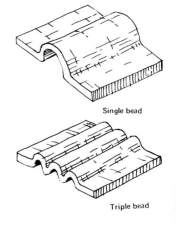

Single bead

Triple bead

trimming a bead

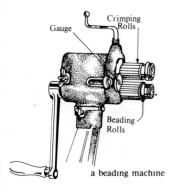

a beading machine

Crimp

Meaning

A bead is a narrow, half-round moulding in metal. Cylindrical metal containers which have to be lifted, such as a refuse bin, usually have beading on their top edge, as a sharp edge there could be dangerous. The forming of beads on sheetmetal to increase the strength of a metal object (or to ornament it) is called **swaging**. A welder is said to *run a bead* when a weld is made along a joint or along another weld.

A bead is also a small convex moulding made in wood, plastic or other material. A **beaded-joint** is a joint in which one of the butting edges has a bead along its edge. A **bead plane** is a special plane for cutting beads from solid materials or for making grooves into which beading can be placed.

Associations

See: **joint, moulding, plane, swaging, weld.**

Bevel

Pronounced: BEVAL (*e as in pet, a as in ago. The word rhymes with level*)

Origin

From the Old French *baif*, which became *baivel* meaning *with a gaping, open mouth*. The notion of **slanting** developed from this position of the mouth.

Meaning

A bevel describes a slanting surface which slopes from the horizontal to the vertical and is not a right angle.

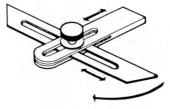

an engineer's bevel

a bevel on a chisel

For example, a chisel has a cutting edge which is shaped with a bevel. When the angle is 45°, the bevel is usually called a **mitre**.

A *bevel* is also a tool (something like a T-square) with

an adjustable blade, which can be locked into a desired angle with a set screw. Usually called an **engineer's bevel**, it comprises a blade, (cut at one end at 45° and 60° at the other) and a stock which can be adjusted in relations to each other. It is used for duplicating, testing and setting out angles.

In welding, bevel describes the angle cut on the edge of a metal plate preparatory to the welding process.

Associations

The bevelled cutting edge of a tool is called the **bezel**. **To splay** is to make something bevelled.
See: **angle, arris, camber, chamfer, chisel, facet, honing, mitre, rake, scarfing, weld**

planing a bevel

Bit

Pronounced: BIT (*i as sit*)

Origin

From the Old English *bita* meaning a *bite* or *a piece bitten off*. Originally, *bite* meant to *split* or *cleave*, which relates to the meaning of *bit* as used in a drill.

Meaning

A bit is the cutting iron of a plane, the head or cutting edge of an axe, and the head of a soldering iron.

There are many kinds of brace bits and drill bits, which are used with power or hand drills, each of which has a specific boring function, including: twist bit, centre bit, countersink bit, dowel bit, expansive bit, masonry bit, Forstner bit, spade bit (also called a flat power bit), pin bit, spoon bit, Swiss twist bit, screw bit, combination bit, hole bit and auger or twist bit. A **rosebit** is solid and cylindrical and is used for countersinking drilled holes.

Bits usually have a square-shaped, tapered shank or **tang**, which will fit tightly into the jaws of a brace. The cutting edges of a bit are called the **lips**. A bit is usually grooved (See: **fluted**), in order that metal cuttings can be directed away from the hole so that it is not clogged up.

Depth stops can made or purchased which can be attached to bits so that a hole may be bored to exactly the depth required. An **extension bit** has a long shank,

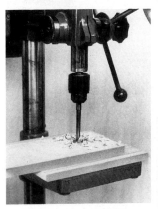

using a forstner bit

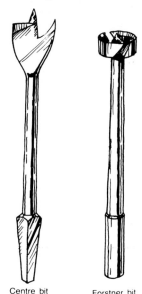

Centre bit   Forstner bit

17

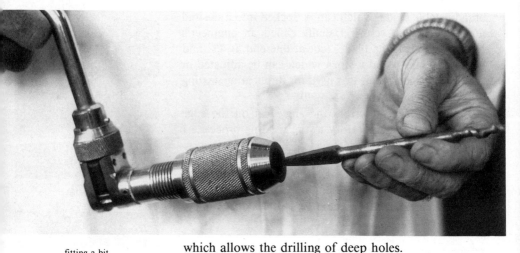

fitting a bit into a brace

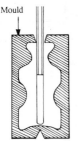

Bradawl

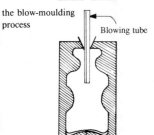

the blow-moulding process

Mould

Blowing tube

which allows the drilling of deep holes.

The term **bit** is also an abbreviation of *binary digit*, which is the smallest unit of storage capacity of a computer. One bit is sufficient to tell the difference between *yes* and *no*, *up* or *down*, in fact any two opposites.

Associations

A **bradawl** is often used to start a hole before a drill and bit are used. A **collet** is the part of a drill which opens and closes to fit the shanks of different-sized bits and other drill accessories.

See: **boring, brace, chuck, computer graphics, countersink, dowel, drill, fit, fixing, plane, rout, solder, stop.**

Blow Moulding

Pronounced: BLO-MOLDING (*o's as in go*)

Origin

Blow is derived from Old English *blawan* meaning *to puff*. Moulding is from the Latin *modulus* and later Old French *modle* meaning *form* or *pattern*. Note the similarity to the modern word *model*.

Meaning

Blow-moulding is a method of mass-producing hollow objects (such as bottles) by blowing compressed air into a cylinder of soft thermoplastic polymer placed in a cold mould, so that the plastic expands to create the shape of the mould. Commercially, there are a number of methods of blow moulding, the most common methods being **extrusion blow moulding** and **injection blow moulding**.

Associations

See: **extrusion, die, injection moulding, moulding, plastic, shape, thermoplastic, vacuum forming**

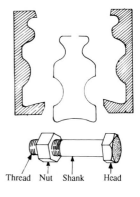

Bolt

Pronounced: BOLT (*o as in old*)

Thread Nut Shank Head

a nut and bolt

Origin

From the Old English *bolt* meaning a *bar (of iron or wood)*, usually used to fasten a door. From this the term *bolt* came to mean *to close (a door)*, or *to fasten together (with bolts)*.

Types of bolts

Meaning

A **bolt** is part of a door or gate fastening consisting of a sliding metal bar (the bolt) on the door and a socket on the jamb (the side-post of the door) or lintel (horizontal timber over a door). It is also a metal pin with a head (which vary in shape), which is used for holding things together (especially large sections of timber, heavy wooden objects and fairly thick metal sheets). Very often it is used with a **nut**, which is a block of metal which is bored with an internal thread which receives a bolt. Bolts are usually described by their length (i.e. of the shank in millimetres), shape, head form and thread form (i.e ISO metric thread diameter) and the metal from which they are made. Bolts are often used when fixing timber to concrete or masonry. Two bolts which are commonly used in brick are the **loxin** and the **dynabolt**. These bolts are sometimes termed **masonry anchors**.

square   hexagonal

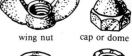

jam   castellated

wing nut   cap or dome

thumb or knurled   stop or nyloc

types of nuts

Associations

A **stud** is similar to a bolt but is threaded at both ends. A **U bolt** is shaped like the letter U. A **cotter pin** (a form of split-pin) is sometimes inserted into a hole near the end of a bolt to prevent an attached nut from working loose.

See: **clearance, counterbore, fasteners, fixing, drill, scarfing, screw, rivet, wrench**.

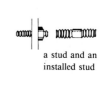

a stud and an installed stud

cotter pin

Boring

Pronounced: BORING (*o as in story*)

Origin

From Old English *borian* meaning *to cut with a sharp point, to pierce*. The **auger**, a simple boring tool, was used in Ancient Egypt before 600 B.C.. The first machine for boring was made to bore a cannon in 1540 by an Italian, Beringuccio. An Englishman, John Wilkinson, invented a metal boring machine in 1775 and the first worm-driven cylinder boring machine was invented in 1799 by an Englishman, William Murdock.

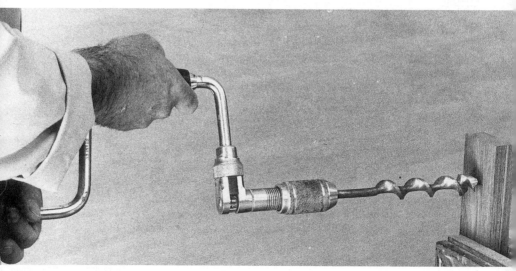

boring a hole with a brace and bit

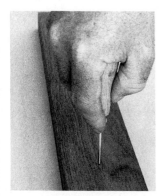

using a bradawl

Meaning

Boring is a process of making a hole in some material, using a revolving cutting tool which usually has a sharp point. The hole is usually hollowed out evenly. The term **boring** is generally used for making holes in wood; **drilling** is used for making holes in metal or masonry. A **boring mill** is a machine which enlarges holes to a finished size required, usually by using a single cutting tool.

A **bore** is the internal diameter of a hole, pipe or cylinder

Associations

A **bradawl** and a **gimlet** are small wood-boring tools, used for boring small holes ready to accept nails or screws. A **reamer** is a boring tool used to finish a hole

accurately and with a smooth surface.
See: **bit, brace, chuck, coolant, drill, jig, lap, lathe, mandrel**.

Brace

Pronounced: BRAS (*a as in face*)

Origin

From Old French *brasse* meaning *the two arms* and *bracier* meaning *to embrace (with both arms)* and *clasp and hold tightly*.
The Ancient Romans used a revolving tool called an auger for drilling. It comprised a curved blade which was turned by a crossbar. It was the forerunner of the drilling tool, a *brace*, which was the first complete **crank**, where linear motion was converted to rotary motion. It was invented by a Flemish carpenter about 1400.

Meaning

A **brace** is a strengthening piece of timber or metal which holds timber or metal firmly together, usually to support a structure. A **brace** is also a revolving tool for boring (as in **brace and bit**).
To brace something is to secure, support, fasten or tighten it.

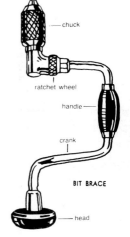

Associations

See: **bit, boring, chuck, cleat, drill, fasteners**.

Brazing

Pronounced: BRA-ZING (*a as in face, i as in ring*)

Origin

From the French *braser* meaning *to solder*. A *braise* was a glowing charcoal and a *brazier* was a pan for holding hot coals. Brazing also derives from Old English *braesian* (from *braes* meaning *brass*) which means to cover with brass.

Meaning

Brazing is an oxy-acetylene welding process of bonding

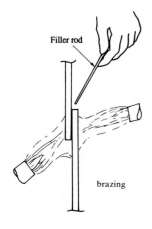

brazing

the surface of two metals together using a non-ferrous filler-rod alloy (often bronze), which has a melting point above 800°F.. The rod is melted and directed so that the molten metal flows by capillary action between the closely-fitted surfaces of the two metals. **Capillary action** is caused by surface tension, adhesion and cohesion forces in the filler material which cause the liquid produced to flow evenly between closely-fitted surfaces. In this **braze welding** process the base metal pieces to be joined do not melt.

Associations

See: **alloy, filler, scarfing, weld**

Bromide

Pronounced: BRO-MIDE (*o as in go, i as in side*)

Origin

From the Greek *bromos* meaning *a bad smell*, then applied to bromiate, the salt of bromic acid. The element *bromine* was discovered in salt water by Antoine Jerome Balard in 1826. Johann Heinrich Schulze, a professor of Anatomical Science at the University of Altdorf in Germany in the early 1700s, was the first to discover that a compound of nitrate (silver nitrate) blackened when exposed to bright sunlight and could produce an image. This was the beginning of the use of light-sensitive materials, essential in photography.

Meaning

The word bromide is often used for bromide paper, which is a photographic printing paper coated with a silver bromide emulsion which makes the paper very light-sensitive. The emulsion is held to a base of plastic or paper by **gelatin**. In the presence of light, silver bromide decomposes into metallic silver and bromine and appears black. Bromide paper is the most popular form of photographic enlarging paper because it needs quite short exposure time to give good black images. Owing to its speed, it must be processed in a yellow-orange light. It is made in a number of different surfaces, including glossy, velvet, eggshell, fine and coarse lustre, linen and rough and smooth matt finishes, in several degrees of contrast and in single and double-weight thickness. It is usually developed in about a

a diffusion transfer process to produce bromides for printing reproductions

photograph by kind permission of Agfa-Gevaert

minute in normal strength solution.
Regular bromide papers are blue-sensitive only and are used for most black and white printing. **Panchromatic bromide** is sensitive to all colours of light. It gives clear black and white prints from colour originals where blue-sensitive paper would produce distortions. **Chlorobromide** which has emulsions with both silver bromide and silver chloride produces a rich brown-black image.

Associations

Sensitometry is the scientific study of light-sensitive materials and their behaviour.
See: **emulsion, exposure, filter, fixing, grain, half-tone, negative, positive, texture**

Buffing

Pronounced:BUFING (*u as in stuff*)

Origin

From French *buff* meaning *a buffalo* and *the skin of a buffalo*. The word came to mean *to make as smooth as buffalo leather.*

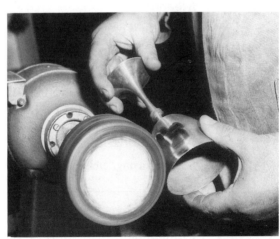

buffing

Meaning

Buffing is the polishing or sharpening of metal objects and tools and also the polishing of the surface of plastic objects. Usually a buffing **mop** or **bob** is a piece of leather coated with a **buffing compound**, such as tallow sprinkled with fine emery powder. Valve-grinding paste spread on leather is also an effective buff for metal.

A **buffing machine** is a powered polishing-machine, which usually has a calico mop, impregnated with a cutting or polishing compound, at each side. Scratches can be removed from metal and surfaces polished to a high lustre by pressing the metal against one of the revolving buffing mops.

In photography a **buffer** is a chemical which retards the chemical reactions in the solution being used.

Associations

Tripoli is a buffing compound which comprises both an abrasive and a lubricant. It is used to produce a lustrous finish on the surface of metals. It is usually manufacture in stick form.

See: **abrasive, accelerator, finish, grind, lap, plastic**

Burin

Pronounced: BUR-IN (*u as in pure, i as in win*)

Origin

From *burin* the French word for *a graving tool* or *etcher's needle*, which derived from the Italian *burino*.

Meaning

The burin is probably the most useful tool in engraving. It is made of a metal rod with a pointed triangular head of tempered steel and usually has a wooden handle. It is used to scoop out wood, metal or other material. A burin is sometimes called a **graver**.

A burin is also used in photography for retouching etched plates.

Associations

See: **engraving, etching, temper.**

a burin

Burnish

Pronounced: BUR-NISH (*u as in fur, i as in dish*)

Origin

From an Old French word *burnir* meaning *to brown*.

Meaning

To burnish is to polish something (e.g. clay, glass, metal) by rubbing it (without the use of an abrasive)

burnishing

with a smooth metal object to create friction. A **burnisher** is a tool of hardened steel which is used to give the final polish to metals, using friction. Sometimes metal is burnished by *tumbling* the metal with steel balls which move at high speed.

Associations
See: **finish**.

Burr

Pronounced:BUR (*u as in fur*)

Origin
The origin of burr is not known but it was used in the past for a sharp. prickly, clinging seed-vessel of certain plants. They were quite dangerous to touch.

Meaning
A burr is a sharp, ragged, projecting edge which often remains on metal after it has been cut, stamped, machined or drilled. The burr can readily cause an accident if not removed.
Burring is the turning of the edges of cylindrical or curved metal objects to produce **flanges**. It can be done manually using a variety of **stakes** or by using a machine called a **burring machine** or a **jenny**.
A burr is also a small, rotating cutting tool. It is used. for example in drilling in dentistry and in engraving work.

Associations
See: **flange, joint, stake, swage.**

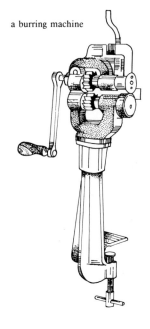

a burring machine

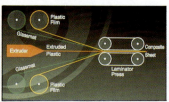

calendering

All width sheet Calender Rolls

Calender

Pronounced: KALANDA (*1st and 2nd a's as in ant, final a as in ago*).

Origin

From the French *calandre* meaning *a roller* or *mangle*

Meaning

A calender is a machine with metal rollers. When material is passed through the rollers, it is pressed, smoothed and sometimes glazed. **Calendering** in plastics is when heated thermoplastic material is passed between two or more calenders to emerge as a flat film or sheet. Calenders are used to coat sheets of materials, such as fabrics, card and paper.

Associations

See: **glaze, thermoplastic**.

Callipers

(*Callipers may also be spelled calipers*)

Pronounced: KAL-IP-AS (*1st a as in cat, i as in ship, 2nd a as in ago*)

Origin

Callipers are a variation of *calibre* (or *caliber*) from the Italian *calibro* which comes from the Arab word *qalib* meaning *a mould for casting metal*. Calliper compasses were used for measuring the calibre (diameter) of a bullet. Bronze outside callipers were used in Pompeii in A.D. 79. Vernier callipers were invented in 1851 by J.R.Brown of the United States.

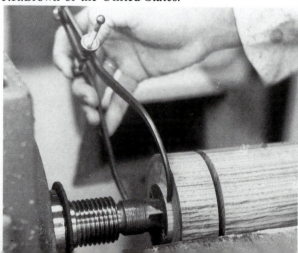

Meaning

Callipers are tools made of wood, metal or plastic which are usually used for measuring convex, concave or irregular-shaped objects. **Inside callipers** measure the diameter of holes and cylinders; **outside callipers** measure the outside diameter of cylindrical objects. They are used to measure the thickness of a sheet of paper or board or film in microns (millionths of a metre). Both kinds can be used for measuring distances over or between surfaces. They are not direct-reading tools. Calliper settings are measured using a scale or a **micrometer**. Work which requires a degree of accuracy too fine to be measured by callipers or rules are measured by a **micrometer**.

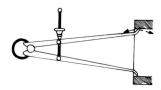

checking a diameter using inside callipers

Associations

A **Vernier calliper** is a precision tool which is able to measure as accurately as 0.025 mm. It is usually used where minute tolerances are required. A **hermaphrodite calliper** (also called a *morphy*), which has one sharp, pointed leg, is used for special lay-out work. **To calibrate** means to find the **calibre** of something.
See: **elevation, proportion, scale, tolerance.**

measuring an internal diameter with a vernier calliper

Cam

Pronounced: KAM (*a as in lamb,*)

Origin

From the Dutch *kam* meaning a *comb* or something *tooth-shaped.*

The cam is believed to have been invented by a methematician called Hero, who was born in Alexandria in Egypt in the first half of the 2nd century. Cams were important parts of water wheels which controlled heavy hammers used in ore crushing.

Meaning

A cam is a piece of metal or plastic which is so shaped that it can convert circular motion into reciprocal (alternating backward and forward) or variable motion. It can either rotate or slide. It is part of a mechanical device where there is direct contact between the edge (or *profile*) of a cam and another part of the mechanism called the **follower**, which may move back and forward

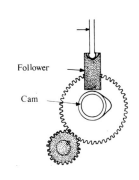

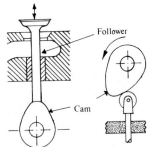

(reciprocally) or oscillate about a fixed axis. The shape of a cam (e.g. pear-shaped, circular, heart-shaped) controls the motion of a follower, which vary in design.

A cam is found in car engines (on a **camshaft**, which operates the valves of an engine), locking devices, textile machines, sewing machines, printing machines and many automatic machine tools.

Associations

See: **machine, reciprocate**.

Capacitor

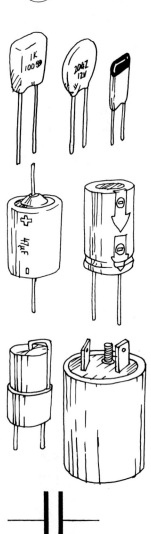

capacitor symbol

Pronounced: KAPASI-TA (*1st and 2nd a's as in cat, i as in sit, 3rd a as in ago*)

Origin

Capacity is from the Latin *capacitas* meaning *the ability to receive or contain*. From this term the word *capacitance* developed (when electricity was discovered) meaning *an ability to store a charge of electricity*. The term **capacitor** was coined to describe the device which had capacitance.

Meaning

A **capacitor**, sometimes called a **condenser**, is a term in electricity to describe a device used for storing (or *condensing*) an electrical charge. It consists of two conducting plates separated by an insulating material called the **dialectric**, which is a **non-conductor**. They are made from various materials and are available in a variety of shapes and sizes. They are classified according to the insulating material used in the dialectric, e.g. ceramics, glass, plastic (e.g. polyester or polystyrene film), metallised film or paper saturated in oil. The **capacitance** of a capacitor is measured in **farads**. **Fixed capacitors** have a specified capacitance; **variable capacitors** have capacitance which may be changed or varied. They perform a variety of functions, such as smoothing the flow of fluctuating electrical current, blocking a continuous flow of direct current but allowing an alternating current to flow. Almost all electronic devices use capacitors.

Associations

See: **conductor**

Carcass or Carcase

Pronounced: KAR-KAS (*1st a as in cat, 2nd a as in ago*)

Origin

From the French *carcasse* meaning the *dead body of an animal or bird* or *the bones of a cooked bird*.

Meaning

A carcass is the frame of a house or of a piece of furniture. For example, the carcass of a chest of drawers under construction comprises the framework of the box-like piece of furniture before its sides, doors, shelves and any decorative features are added.

Associations

The horizontal **members** of a frame are called **rails**, and the vertical members are called **stiles**.
See: **fasteners, furniture, hardboard, joint.**

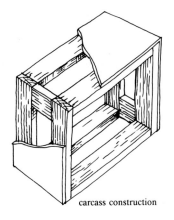
carcass construction

Casting

Pronounced: KAST-ING (*a as in car*)

Origin

Cast is from the Old Norse *Kasta* meaning *to throw*. The Ancient Egyptians cast bronze in moulds over 3500 years ago.

Meaning

To cast is to create an object (called a **cast** or **casting**) by putting (that is to cast) molten metal or plastic material (e.g. resin) into a mould (also spelled *mold*), in order that it will harden into a required shape. **Crucible furnaces** are used for melting metals which are transferred in a **crucible** into a mould.
In **sand mould casting**, a **moulding box** or **moulding flask** is used, whose upper half is called the **cope** and whose lower half is called **the drag**. In the moulding flask is tampered and dampened sand (called the **core**), into which is placed a **pattern**, usually made of wood but sometimes of styrofoam. The sand is **tamped** (That is rammed or packed.) around the pattern (using a **rammer**) and then the pattern is removed leaving a cavity, which is the mould into which the molten metal

a casting flask

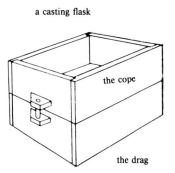
the cope

the drag

a rammer for ramming sand around a pattern in casting

is poured through a hole in the sand (a funnel-shaped opening where the thick, top part is called a **gate** and the thin, bottom part a **sprue**) to create a metal object like the pattern.

Plaster mould casting uses plaster moulds instead of sand moulds. This process gives a better surface finish to objects than those cast in sand.

Slush casting is a process for the production of plastic objects. A mould is heated in an oven and then the mould is filled with plastisol, which is slushed about inside the mould to produce the required thickness. The mould is reheated to harden the plastisol and it is then cooled. The finished object is then removed from the mould.

Associations

Foundry sand is a mixture of silica and fine sand, which does not dry out and can be used constantly. It can be readily shaped to produce precise and detailed castings.
To strickle means to level off the sand in a moulding box when preparing moulds for casting.
See: **clearance, cove, crucible, cire perdu, draft, drawing, moulding, pattern, sculpture, warp.**

Catalyst

Pronounced: KATA-LIST (*1st a as in cat, 2nd a as in ago, i as in big*)

Origin

From the Greek *khatalusis* meaning *dissolving* or *loosening*. The term *catalyst* was first used by a Swedish chemist, J.J. Berzelius (1779 - 1862) who discovered **catalytic action**

Meaning

A catalyst is a substance which accelerates the rate of reaction of substances yet remains unchanged itself at the end of the reaction. This process is called **catalysis**. The reason a catalyst changes the speed of reaction of substances is that the new reaction requires less energy than the reaction without a catalyst. There are two kinds of catalysts: those which form a suitable surface for a reaction of the substances to take place and those which take part in the reaction but which are reformed on the completion of the reaction sequence. A catalyst is needed to initiate the setting (curing and hardening) of a synthetic resin. A catalyst helps a **monomer** to change to a **polymer**.

Associations

Biological catalysts are called **enzymes**.
See: **accelerator, anodise, curing, lacquer, particleboard, plastics, resin.**

Chamfer

Pronounced: CHAM FA (*a as in lamb, a as in ago*).

Origin

From the Old French *chanfraidre* meaning *to bevel off* or *to break the edge.*

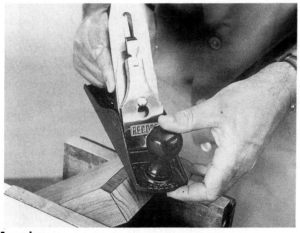

planing a chamfer on an end grain

Meaning

To chamfer is to remove the sharp edge or corner of a piece of wood (See **arris**) by planing the edge at an angle to produce a **bevel**. The angle is generally 45°. A chamfer is usually made for decorative purposes but is sometimes made on plinths, skirting-boards and some mouldings to decrease dust build-up. When a surface is planed to an angle of more than 45°, it is called a **splayed edge**.

A **chamfer plane** is made with adjustable guides, so that a range of chamfers can be cut.

planing a splay

Associations

See: **angle, arris, bevel, countersink, moulding, plinth, rake, rout**

Chasing

Pronounced: CHAS-ING (*ch as in church a as in late*)

Origin

From the French *enchasser* meaning *to emboss, engrave* or *encase and frame*

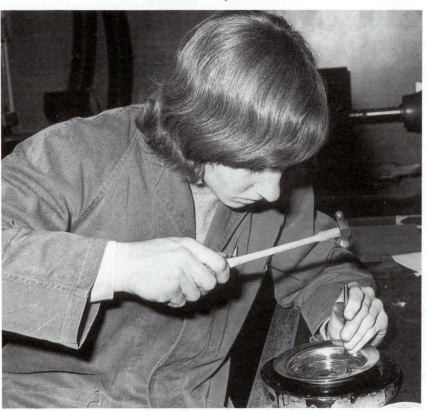

chasing

chased jewellery

Meaning

Chasing is to decorate a metal object (such as a sword) by indenting its surface from the front (See **Intaglio**) with a hammer, punch, or some tool without a cutting edge to form a design. Chasing tools (e.g. hammers) have heads which can produce patterns showing ovals, squares, domes and lines. The word also applies to the removal of roughness or blemishes on metal which is being modelled, e.g. as in **cire perdu** casting.

Chasing is also the following of an outline or design on sheetmetal with a *chaser* or *liner*, which is a punch shaped at the end like a screwdriver. It is usually used

with a repoussé hammer.
It also means a lengthwise groove which receives a male part to form a joint.

Associations

A **chaser** is also a lathe tool with a cutting edge in the form of a screw. It is used to finish accurately screw threads cut on a lathe.
See: **cire perdu, dapping, die, emboss, impression, intaglio, joint, relief, repoussé, scoring, tooling**.

different chasing hammmer heads

repoussé or chasing hammer

Chisel

Pronounced: CHIZAL (*ch as in church, i as in pin, a as in ago*)

Origin

From the Old Norman French *chisel*, which came from the Latin *cisellum* meaning *an instrument for cutting*. Chisels and gouges have been used for thousands of years, as they were used in Ancient Egypt, Rome and Greece.

Meaning

A chisel is a tool which has a steel blade with a square, bevelled edge at one end, which is used for shaping wood, stone, metal or plastic. At the non-cutting end of the blade is either a square tapered tang which fits into a wood or plastic handle (a **tanged chisel**) or a forged socket into which a tapered handle fits (a **socket chisel**). Chisels are classified into three groups: **firmer chisels**, **paring chisels** and **mortise chisels**. A **cold chisel** is a used for cutting materials (such as concrete, pipes, rivets, etc.), for chipping metal and for splitting nuts which have become rusted and cannot be loosened.
To chisel is to cut or shape with a chisel.

a firmer

using a gouge

Gouges are a type of chisel with a concave blade which is ground and sharpened on the inside or outside surface, depending upon the kind of carving or paring work to be undertaken.

There is a very wide range of gouges with different curvatures to the blades (designated by numbers for size), which are used mainly for wood carving and wood sculpture.

Associations

See: **bevel, contour, engraving, groove, lathe, sculpture, shape, turn.**

Chuck

Pronounced: CHUK (*u as in luck*)

Origin

From the Old French *choque* meaning a *block of wood* or *wedge*. It was a wedge used particularly for stopping the movement of a wine cask or the wheel of a cart. *To chuck* or *chock* also meant *to close tightly* or *to close up*.

Meaning

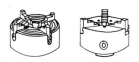

4-jaw independent chuck

drill chuck

A chuck is a device for holding a piece of work or a cutting tool on a lathe. There are various different chucks, which are used for specific purposes. Most chucks have 3 or 4 adjustable jaws (called **dogs**) which can be tightened to hold the work or tool. A **universal chuck** has jaws which move simultaneously and regularly to tighten on a hexagonal stock. An **independent chuck** has four reversible jaws, each of which can be adjusted independently. It can be used to hold pieces of work of almost any shape and because its jaws can be adjusted independently, great accuracy can be achieved in the setting of a piece of work.

A chuck is also the part of a carpenter's **brace** which holds a **bit**.

Associations

See: **bit, boring, brace, drill, lathe, rout.**

Circuit

Pronounced: SER-KIT (*er as in herd, i as in fit*)

Origin

From the Latin *circuitus* meaning *going round* from *circum* meaning *round* and *it* from *ire* meaning *to go*. Compare with *circular, circulate, circumference*.
See: **electrode**.

Meaning

A circuit is the path of an electrical current (*the electron flow*) through electrical components and back to a power source. A **closed circuit** or **complete circuit** is when an electrical flow is continuous and the circuit operates. An **open circuit** or **broken circuit** describes the situation when the electrical flow stops (either deliberately by means of a switch or accidentally because of, for example, a broken wire) and the circuit does not operate. A **series circuit** is one in which all of the electricity must pass through all of the components of the circuit. If a component is removed from the series, the circuit does not operate. In a **parallel circuit** there is more than one path for the electricity to flow along. The electricity can be divided between each parallel branch in the circuit. A **short circuit** is when electricity is able to take another, lower resistance path than that when it flows through a normal circuit. When this happens the connecting wires in the shortened circuit may become hot and melt.

A **circuit breaker** is a safety device to cut out an electrical circuit when it is overloaded. Unlike a fuse which has to be replaced, a circuit breaker can be reset after a fault has been dealt with. A **circuit diagram** is a diagram which shows exactly how components are connected in an electrical circuit. The lines representing wire are always straight or with right-angle bends. Each component in a circuit is identified by its own graphic symbol.

Associations

An **integrated circuit** is a micro-circuit, containing thousands of transistors, diodes, resistors and capacitors, fashioned within a single crystal made of thin wafers of silicon called a **silicon chip**, which is only a few millimetres square. A **diode** is a **semi-conductor** in that

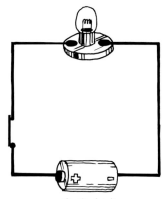

a simple circuit.

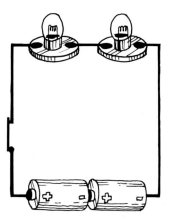

a series circuit

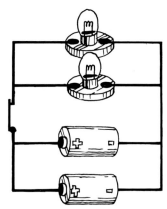

a parallel circuit

it is a device which allows an electric current to flow through it in only one direction.
See: **angle, current, electrode, electronics, fusability, key, transistor.**

Cire Perdu

Pronounced: SEAR (*ea as in fear*), PAIR-DU (*ai as in fair, u as in due*)

Origin

From the French *cire* meaning *wax* and *perdu* meaning *lost*. The cire perdu (lost wax) process of casting was used in Egypt in 2500 B.C..

Meaning

Cire perdu is a casting technique where a model is made first in wax. The wax model is covered in soft clay and plaster to form a mould when dry. The wax is melted and drained from the plaster mould and then molten metal is poured into the mould to produce a replica of the original design. Another technique is where the metal is forced into the mould by centrifugal force.

The cire perdu process is more accurate and produces a better surface finish than sand casting. It is possible to reach an accuracy of up to 0.05 mm.

Associations

Cire perdu is also called the **lost wax method** or **centrifugal casting**.

See: **casting, chasing, filigree, moulding, replica, sculpture.**

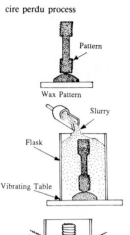

cire perdu process

Pattern
Wax Pattern
Slurry
Flask
Vibrating Table

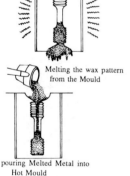

Melting the wax pattern from the Mould

pouring Melted Metal into Hot Mould

Cladding

Pronounced: KLAD-ING (*a as in glad*)

Origin

From Middle English *cladian* meaning *to clothe* or *cover with clothes.*

Meaning

Cladding is the coating, sheathing or covering on a structure or a material for a number of reasons, such as to give protection against the weather, provide thermal or accoustic insulation, to prevent scratching or for decorative effects. In building construction, it is the external covering to a timber-framed or metal-framed building, such as timber weatherboards or panels of aluminium or galvanised iron with a zinc anneal, which are usually coated with a plastic enamel.

Associations
See: **anneal, enamel, plastic**

Clearance

Pronounced: KLEAR-ANS (*ea as in fear, a as in ago*)

Origin

Clear is from the Latin *clarus* meaning *plain, clear, manifest. Clearance* is a modern derivation from clear.

Meaning

Clearance refers to the *clear* space allowed for the passage or movement of two adjacent objects or parts or between moving and stationary parts of a machine. It is also the angle and space between the cutting edge of a lathe tool or other cutting tool and the vertical position of the workpiece.

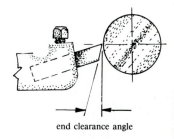

end clearance angle

Associations

Draft refers to the clearance on a pattern which permits the easy withdrawal of a pattern from a mould after casting. A **clearance hole** is a hole slightly larger than the diameter of a stud or bolt it will receive, so that assembly is made easy. **Shims** are pieces of sheet metal or steel of various sizes which are placed between mating parts to make a required clearance.
See: **bolt, casting, draft, fit, lathe, moulding, pattern, relief, saw, spline, tolerance**

Cleat

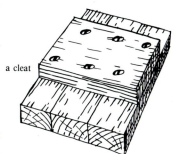

a cleat

Pronounced: KLEET (*ee as in see*)

Origin

From Old English *cleat* meaning *a wedge* or *clod* or *a round lump*. It came to mean a piece of wood or metal,

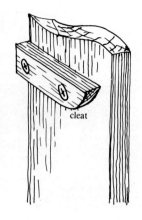
cleat

especially one that was bolted on to another material to secure ropes (e.g. on ships) or to strengthen woodwork.

Meaning

A cleat is a piece of wood or metal which supports a shelf, beam, bracket or table-top. It also refers to a narrow board or batten of wood which is fixed with nails or screws to a wide board to prevent movement and possible warping of the board. It is also a strip of wood or metal which is fixed to a wall so that other things may be fastened to it, or to locate another piece in its correct position.

Associations

See: **brace, fixing, warp.**

Cleavage

Pronounced: KLEE-VIJ (*ee as in see, ij as idge in bridge*)

Origin

From Old English *cleofian* meaning *to divide, split, chop, peal and strip.*

Meaning

Cleavage refers to the way in which things tend to split. Woods which cleave or split easily in the direction of their grain are said to be **fissile**. A timber which has **cleavage strength** is not easily split with the grain.

Associations

See: **grain**

Colour

Pronounced: KULA (*u as in dug, a as in ago*)

Origin

From the Latin *colare* meaning *to colour*. In 1672, Isaac Newton reported to the Royal Society in England that daylight (white light) when it passed through a **prism** splits into seven colours, **the spectrum**: red, orange, yellow, green, blue, indigo, and violet. Objects, he said, achieve their colour by absorbing or subtracting certain parts of the spectrum and reflecting or transmitting the parts left.

Christiaan Huygens (1629 - 1695), a Dutch scientist, formulated in 1690 the laws of reflection and refraction.

No one knows exactly how colour terms developed. Different people living in different environments use different colour words. Eskimos have many colours to describe snow; desert people have many colour words for yellow and brown. The names of many colours are derived from specific parts of our environment. Derived from plants are apricot, lemon, grass-green, hazel, rose-red. Derived from minerals and metals are alabaster, amethyst, copper, turquoise-blue. Derived from man-made products are chocolate, wine-red, bottle-green. Derived from fauna are beaver, canary-yellow, mouse-grey, fox, butterfly-blue. Derived from geographic names are Berlin-blue, Copenhagen-blue, Naples-yellow, Spanish-green. Derived from natural phenomena are aurora, spring-green, sky-blue, fire-red.

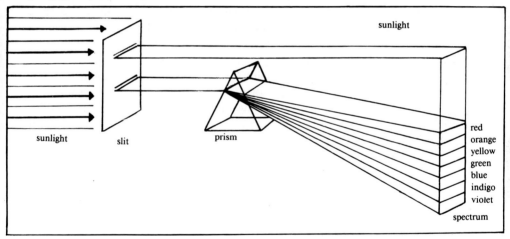

Meaning

Light is the only source of colour and without it no colour can exist. All things are reflectors, absorbers or transmitters of one or more of the colours which make up white light. For there to be colour, three elements are needed: light (the source of colour), the object (and its response to colour) and the eye, the perceiver of colour. Every colour has three characteristics:

1 **Hue** is the quality which distinguishes one colour from another. That is, for example, the yellowness or

Red results from the subtraction of green and blue wave lengths from white light.

Blue results from the subtraction of green and red wave lengths from white light.

Green results from the subtraction of blue and red wave lengths from white light.

Yellow results from the subtraction of blue only, leaving red and green.

blueness of something. There are said to be 150 differences of hue.

2 **Tone** (called also **brightness**) is the lightness or darkness of a colour. All colours have a place on a black-grey-white **tonal scale**. Yellow is close to white, violet is close to black, etc. **High-key** colours fit into the top of the range of a tonal scale; **Low-key** colours fit into the bottom of the tonal scale.

3 **Saturation** (called also **chroma** or **intensity**) is the purity, brilliance or richness of a colour and its strength or weakness (i.e. *high* or *low* saturation). The Latin word *satur* from which saturation comes, means *full*. It refers to the blueness of a blue or the greenness of a green. For example, orange has high intensity; brown has low intensity. One measure of intensity is how little white, black or grey is in the colour.

The sum of hue, tone and intensity is sometimes referred to as the colour **value**. Each colour has its **complementary colour** (see **Colour Wheel**). Coloured lights are termed **transmitted colour** and the colour in painting is called **reflected colour**. A red wall will absorb all colours except red which it reflects.

Primary colours (blue, red, yellow **in pigments** and red, green and violet **in lights**) are colours which cannot be made by mixing other colours. They can be used in different combinations to produce other colours. Two primary colours when mixed (e.g. red and yellow in equal proportions to produce orange) produce a **secondary colour**. The mixing of a primary colour with a secondary colour or two or three secondary colours produces a **tertiary colour**. Red with violet, for example, produces red-violet; orange, with violet and green produces citrine, olive and russet and all their tones.

Paint colours are mixed by an **additive** or **subtractive** process. Additive colour mixing has a starting point of black to which colours are added to produce the hue required. Subtractive colour mixing has a starting point of white (containing all the colours of the spectrum) to which colours are mixed to subtract the colour required. Paints and inks can be mixed to produce tints as follows: black and red gives brown; brown and white gives chestnut; white, yellow and venetian red gives

buff; yellow and white gives straw colour; black, blue
and white gives pearl grey; lamp black and white gives
lead colour; black and indigo gives silver grey; green
and white gives pea green; light green and black gives
dark green; red, blue and black gives olive; yellow and
red gives orange; carmine and white gives pink; emer-
ald green and white gives brilliant green; blue, white
and lake gives purple; lake, white and vermilion gives
flesh colour; blue and lead colour gives pearl; white
and lake gives rose colour.

In the paint industry, the primary colours are called
yellow, cyan (blue-green) and magenta (blue red). **Colour blindness** is sensitivity to only two distinct colours
rather than the three needed to match all colours in
the spectrum. **Neutral colours** are black, grey and white,
which give the impression of lightness or darkness.
Shade indicates that a colour is changed by the addition
of black to it. **Tint** is when a colour has been slightly
changed by the addition of white to it.

Colours are said to have a **temperature**. **Hot** or **warm**
colours are red, red-orange, red-violet, yellow, yellow-
orange. **Cold** or **cool** colours are blue, violet, yellow-
green, green, blue-green, blue. Cool colours suggest sensations of coolness and seem to *recede* in a painting
and suggest depth; warm colours suggest sensations of
heat and seem to *advance* towards the spectator and
attract. Experiments have shown that there is a difference of 5-7 degrees in a person's feeling of heat or cold
between a room painted in blue-green and one painted
in red-orange. The term **local colours** is used for the
natural colour of an object. For example, a road surface
may be naturally grey but seen in sunlight through
green foliage, it may appear violet. Colours also have
surface characteristics, such as being **opaque** or **transparent** and **gloss** or **matt**.

Associations

Coloration is the arrangement of colours. A **colorimeter**
is an instrument for measuring the intensity of colours.
Chromatic is from the Greek *Khroma* meaning *colour*
and means *full of bright colours*. **Monochromatic** means
a painting in various fine gradations of tones or tints
in one colour, or the use of black and white only.
Achromatic means tones without any hues-being mix-

tures of black and white to produce grey. **Chromatography** is a laboratory, industrial and research technique for the separation, isolation, purification and identification of complex mixtures, which was originally used for the separation of colour compounds. **Earth colours** are pigments, such as brown, yellow, green, red, ochres, ambers etc. which are found naturally in earth and clay.

Colour Wheel

Pronounced: KULA(*u as in bonus, a as in ago*), WEEL(*ee as in see*)

Origin

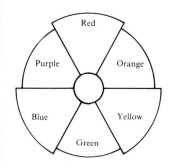

Colour wheels have been devised to attempt to give precise terms for colours. One of the best known and much used, after the language of colour was standardised in the 1930's, was that produced by an American portrait painter, Albert Munsell, in 1915. He devised his system in the form of a tree which classified colours according to the qualities of hue, value, and chroma. In this system each colour can be specified by letters and numbers.

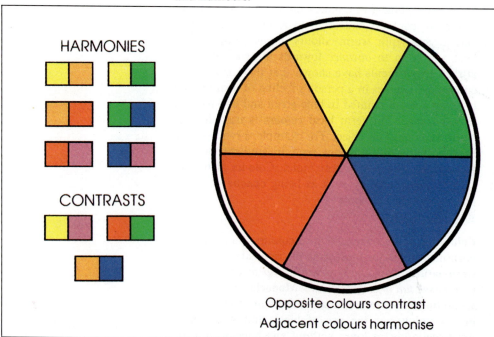

Opposite colours contrast
Adjacent colours harmonise

Meaning

A colour wheel is a wheel (or a circle or chart) which shows the primary colours and mixtures of colours organsied in order of hues and intensity. The colour wheel shows the main gradations of colours.

Complementary colours (note the spelling. It should not be confused with compli*mentary*) are colours which are opposite to each other and the greatest distance apart from each other on a colour wheel. Each **primary colour** (red, blue, yellow) has a **complementary colour** produced by a combination of the other two primary colours. For examples the complement of red is green (i.e.yellow and blue), of yellow is violet, of blue is orange. **Analogous colours** are colours next to each other on the colour wheel. When complementary colours are placed side by side, they intensify each other. For example, red strengthens green.

The colour system most often used is the **Munsell Colour System** which has 1200 samples of colour, but other systems are used, such as those by Ostwald, Prang, Michel-Eugene Chevreul, Harold Kuppers, Frans Gerritsen, and Johannes Itten.

Composition

Pronounced: KOM PA-ZISH-UN (*o as in bomb, a as a in ago, i as in ink, u as in fun*)

Origin

From the Latin *com* meaning *with* and *ponere* meaning *to put*. That is *to put together.*

Meaning

A composition is the combination of elements (colour, shape, size, texture, tone, mass, direction and line) in a work , which, by using the **principles** of balance, rhythm, dominance, contrast, harmony, and unity are harmonized into a unified whole. Each element in a work is important but the whole is more important than the parts.

It refers, too, to the ingredients of some materials, for example, the composition of solder is tin and lead; that of concrete is gravel, sand, cement and water.

Associations

To compose is to arrange things in a specific manner for a particular purpose, as in the positioning of images in a photograph.

See: **contrast, elements, engraving, etching, form, format, hardening, intaglio, motif, pattern, plane, relief.**

Computer Graphics/Art/Design

Pronounced: KOM-PEW-TUR (*o as in lot, ew as in few, ur as in fur*), GRAF-IKS (*a as in bat, i as in ink*), ART (*a as in far*), DE-ZIN (*e as in delay, i as in line*)

Origin

The word computer originates from the Latin *computare* meaning *to calculate* or *count*. The Latin *computus* was a Medievel set of mathematical tables used for calculating astronomical events and moveable dates in the calendar, such as Easter. In Old French the word *comput* was used for the calculation of the date for Easter.

Charles Babbage (1791-1871), an English mathematician, is usually given credit for inventing the computer, as he invented an *electronic calculating and problem-solving machine* in 1834. However, the first machine that stored information for later use was invented by Joseph-Marie Jacquard (1752-1834), a French silk weaver, who used a system in 1804 of punched cards to reproduce woven patterns automatically. The first fully electronic computer was ENIAC (electronic Numeral Intergrator and Calculator) which was invented by J. Presper Eckert and Dr. John W. Mauchly at the University of Pennsylvania in 1945. It used more than 18,000 thermionic valves, weighed 30 tons and occupied 1,500 square feet of floor space. Dr. M.V.Wilkes designed the first all-electric computer in England, which had 18,000 valves, in 1949. *Graphics* is from the Greek *graphe* meaning *writing*. The **binary system** was developed by a German, Gottfried Leibnitz (1646-1716). The terms **computer graphics** and **computer aided graphics** were first used by the Boeing Aircraft Company in the USA in 1900. The first exhibition of Computer Graphics was held in New York, in 1965. The first CAD (Computer Aided Design) systems appeared in an

experimental form in the mid 1960's, and the first commercial drafting and design systems were produced in the late 1960's, mainly to meet the needs of the electronic industry in the designing of printed circuit boards.

Meaning

A computer is an electronic device which can store, retrieve and process data in accordance with a series of stored instructions (*commands*) called a **programme (or program)**. It can make extremely rapid calculations, using either numbers expressed as digits in a given number scale, usually **binary numbers** 0 or 1 (using a **digital computer**) or with numbers represented by measurable quantities of a given magnitude (using an **analogue computer**).

Computer graphics is a technique where programmed data in the form of basic shapes (called **entities** or **primitives**, such as points, lines, vertices, arcs, polygons or circles) are combined to produce **models** (pictorial or geometric). Designs, diagrams, charts and models can be produced. A **light-pen** attached to the computer terminal allows one to draw (or edit) direct onto a computer display screen. The images on the screen can be rotated, enlarged, modified etc.. The same computer equipment can be used for **computer-aided designs** (CAD) and **computer-aided art** (CAA). These are proc-

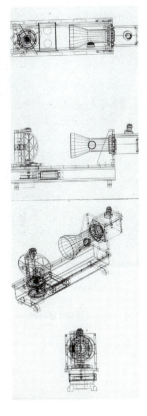

45

esses where a computer is used in the creation or modification of a design or an art work.

Computers can also be programmed to lighten or darken segments of an original image or a photograph to enhance its clarity or intensity. This is sometimes called **computer enhancement**.

inspecting a planar board

Kind permission of IBM Quarterly.

Associations

A **dot matrix printer** is used with a computer to produce graphic data as well as text. Any device attached to a computer (such as printers, plotters or digitisers) is called a **peripheral**. MicroCAD is a term for CAD systems which have recently been developed for microcomputers.

See: **bit, design, develop, electronics, ergonomics, geometric, Industrial Design, pixelated images**.

Conductor

Pronounced: KON-DUKTA (*o as in cross, u as in luck, a as in ago*)

Origin

From the Latin *conducere* meaning *to lead together* and *to guide*
See **electrode**.

Meaning

In electricity, **conduction** means the transmission (the "*guidance*") of electricity by means of a material, which is called a **conductor**. This has the physical properties of being able to conduct electricity from point to point along a circuit with as little loss of energy as possible, because it has what is called **electrical conductivity**. The number of *free electrons* in a material determines its degree of conductivity. Some materials, natural and man-made, have conductivity; some have not. For instance, silver, copper, aluminium, zinc, brass, nickel, steel, tin, lead, carbon and tungsten are conductors. However, air, asbestos, bakelite, ceramics, cotton, ebonite (hard rubber), fibre-glass, mica, oils, porcelain, glass and wood are poor conductors. Materials which are nonconductors of electricity and resist the passage of electricity are called **insulators**, because they *insulate*. That is they prevent the electricity from having any connections with anything else. (The word *insulate* originally meant *to make into an island*, to be separate.) When a current of electricity flows along a circuit through a conductor, some electricity could leak from the conductor. Therefore, the conductor is surrounded by an insulating material to ensure that there is not leakage and *short circuiting*. (see: **circuit**).

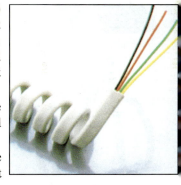

Some substances, such as silicon and germanium, are neither good conductors nor insulators. They are called **semi-conductors**.

Many electrical plugs, sockets, appliance handles are made from phenolic resins, as this plastic is an excellent nonconductor of electricity.

Associations

See: **capacitor, circuit, electrode, electronics, plug, resin, transistor.**

shaping a contour

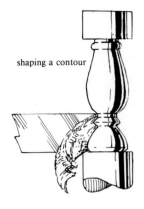

Contour

Pronounced: KON-TOOR (*o as in on, oor as in poor*)

Origin

From the Italian *contorno* meaning *a drawn outline*. Contour drawing was popularised by Kimon Nicolaides in his book "*The Natural Way to Draw*"(1941).

Meaning

A contour is the internal or external edge of a form in a drawing. That is the lines separating an area or shape from its surrounding background. Contour drawing attempts to produce a three dimensional quality and produce a tactile impression as well as a visual stimulus.

Contour also refers to a **profile** or the outline shape of a piece of wood after it has been turned on a lathe.

Associations

See: **chisel, drawing, lathe, line, moulding, profile, shape, texture.**

contrast in line.

contrast in shape.

Contrast

Pronounced: KON-TRAST (*o as in long, a as in fast*)

Origin

From the Latin *contra* meaning *against* and *stare* meaning *to stand*. That is *to oppose, to be different*.

Meaning

To contrast is to set things (**juxtapose** them) in opposition to each other to show striking differences among them.

In photography, contrast is the clear difference between the darkest and lightest parts of a photograph. A **high contrast** photograph has mostly black and white areas with few grey areas. A **low contrast** photograph has mostly grey areas with few black and white areas and is said to be **flat**. **Contrast grades** describes photographic paper which is made in a number of grades, to allow changes in the contrast of photographs. Grade 1 paper is for low (or *soft*) contrast; grade 5 is for high (or *hard*) contrast. A **contrast filter** is a slide of coloured glass which is placed over a camera lens when using black and white film. The filter lightens its own colour and darkens its opposite. A yellow filter, for example, will darken blue sky and lighten white clouds, so that the contrast between the two is heightened.

Associations

See: **composition, exposure, filter, lens, matt, positive, tone.**

contrast in value.

Coolant

Pronounced: KOO-LANT,(*oo as in pool, a as in ago*)

Origin

From Old English *colian* meaning *to grow cold* and *celen* meaning *to make cold*. *Coolant* is a 20th. century derivation from *cool*.

Meaning

A coolant is a liquid which is put onto cutting tools during the machining of metal, in order to reduce the heat caused by friction, to act as a lubricant, to remove swarf and abrasive grains from the cutting area and to keep down the amount of grinding dust. The use of a suitable coolant to reduce heat can increase the working life of a cutting tool and also make possible an increase of 30% in cutting speed. A coolant in metalwork is also called a **cutting compound**.
Soluble oils and mineral oils with water are generally used as coolants in cutting operations. Chemical cutting fluids with water are also used.
The **flood system** is the most common method of applying coolants. The coolant is directed onto a workpiece by a nozzle. The system usually comprises a reservoir, a pump, a filter and a control valve.

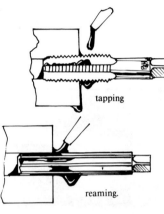

tapping

reaming.

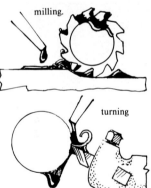

milling.

turning

Associations

See: **boring, drill, machine.**

Core

Pronounced: KOR (*o as in more*).

Origin

From the Latin *cor* meaning the *heart*.

Meaning

A core is the central part of a tree trunk. It is also the middle layer of laminated wood , such as plywood.
In electricity, it is the magnetic path through the centre of a *coil* or *transformer*.
In casting, a core is the mass of sand or other material into which a pattern is placed.

Associations

See: **casting, laminate, medullary, plywood**.

Corrosion

Pronounced: KA-RO-ZHAN (*a's as in ago, o as in rose, zh as s in exposure*)

Origin

From the Latin *corrosionem* meaning *a gnawing to pieces* from the verb *rodere* meaning *to gnaw*. A *rodent* is *an animal which gnaws its food*, and *erode* means *to wear or gnaw away*

Meaning

A metal **corrodes** when it gives up electrons. The metal is usually oxidised (but not always e.g. silver can be corroded by sulphur). That is the oxygen in the air reacts with a number of elements in a fairly complex process of reactions to produce an oxidant, which results in electrons passing from the metal to the oxidant. To prevent corrosion, the loss of electrons from the metal to the oxidant must be stopped. Some of the ways of prevention are to coat the metal with paint, to **galvanise** it (coat it with zinc by dipping it in molten zinc) and to **electroplate** it with chromium, nickel or tin.

Associations

A **caustic** solution is a strong alkali (e.g. caustic soda and caustic potash) which can corrode many other substances.

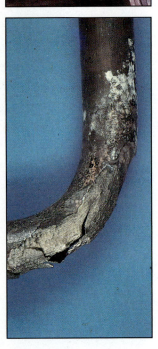

See: **alloy, anodise, electroplating, nails, noble metals, oxidation, paint, priming, preservation**.

Counterbore

Pronounced: KOWNTA-BOR (*ow as in cow, a as in ago, o as in more*)

Origin

Counter is from the Latin *contra* meaning *against, opposite*. Later, it also had the meaning of *different from*. *Bore* is from the Old English *borian* meaning *to cut with a sharp point, to pierce*. To counterbore had the meaning of *boring a different hole from a first hole*.

Meaning

To counterbore is to drill a second hole in wood or metal which is larger than a first hole, using the same centre as the first hole. The second hole is less deep than the first. A bolt or screw can be sunk so that its head does not project above the surface of the wood or metal being worked on.

a counterbored screw

Associations

See: **bolt, drill, screw**.

Countersink

Pronounced: KOWNTA-SINK (*ow as in cow, a as in ago, i as in ink*)

countersinking screws

Origin

From the French *counterfort* meaning a buttress; that is a strong support built against a wall.

Meaning

To countersink is to enlarge a hole with a cone-shaped **bit** sufficiently large to provide a recess for a counter-sunk-head bolt or screw. The hole is **chamfered**. As the screw or bolt is flush or slightly below the surface of the material being worked on, it does not protrude and is therefore "*supported*" (See origin above) by the wood. A **rose head** bit with 8 or more cutters is used to countersink softwood, brass and aluminium; a **snail head** bit with 1 or 2 cutters is used for most hardwoods. Sometimes countersunk holes are used as oil holes.

Associations

See: **bit, bolt, chamfer, screw**.

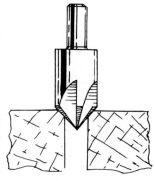

using a countersinking bit

Cramps

Pronounced: KRAMPZ(*a as in lamp*)

Origin

From the Old Dutch *Krampe* meaning *bent*.

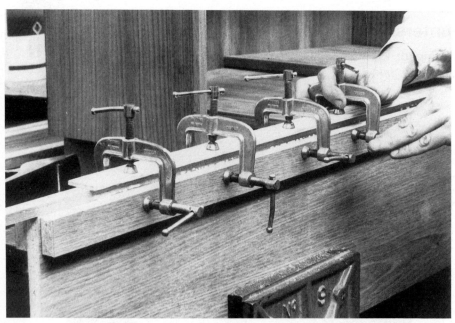

using G cramps

Meaning

Cramps are tools used for holding together things (e.g. lengths of timber, parts of a brick wall under construction, woodwork which has been jointed and glued and is drying, laminated materials which are setting). They are used also to steady wood which is being worked on. There are different kinds of cramps, such as a "G" cramp, an edge cramp, a bar cramp, a pipe cramp, a corner cramp, and a webbing cramp. A **clamp** is the same as a cramp and the two terms are both used.

Cramping time is the time a piece of work should be kept in a cramp to ensure complete bonding, taking into account the adhesive being used, the temperature at the time and the kind of joint being bonded.

using T bar cramps

Associations

See: **adhesive, fastener, fixing, laminate, mitre**.

Crop

Pronounced: KROP (*o as in got*)

Origin

The origin of crop with a meaning of *to cut, prune or trim* is vague but it dates back to the seventeenth century.

Meaning

To crop is to cut off or **mask** out an unwanted area of an illustration or photograph, usually to marks in a margin called **crop marks**. It also means to select a part of a picture or photograph for reproduction.

Associations

A **cropping mask** is a piece of material, usually cardboard, in the shape of the letter L. The overlapping of two cropping L's marks out a part of a work for further use. It is also called **masking** and **blocking**.
See: **shear**.

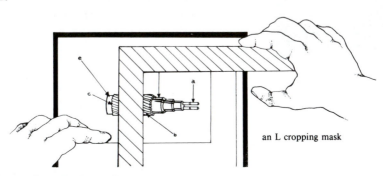

an L cropping mask

Cross Hatching

Pronounced: KROS (*o as in loss*), HAT-CHING (*a as in bat, i as in ring*)

Origin

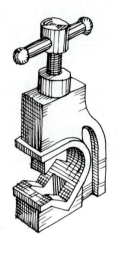

From the French *hacher* meaning either *to chop into small pieces* or *to engrave* or *to draw*. It is related to the French *hachure* meaning *to shade* (usually on a map) to show heights by means of crossed lines. The art was invented by Michael Wohlgemuth of Germany in 1486.

Meaning

Cross hatching is a technique in drawing, etching, engraving, painting and weaving, where parallel lines (straight or curved) are drawn close together (**hatching**) and then other parallel lines criss-cross them at right angles or at an oblique angle (**cross hatching**). The lines produce an effect of shadow, shading or **gradations** of light. Tones can be produced by varying the thickness of the lines.

The term **hatching** rather than **cross hatching** is generally used in mechanical drawing. There is no crossing of lines in a sectioned application.

Associations

See: **drawing, etching, engraving, gradation, half tone, line, mechanical drawing.**

Crosslinking

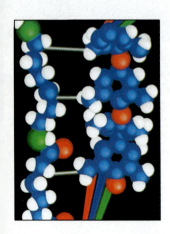

Pronounced: KROS-LINKING *o as in got, i as in bit,*)

Origin

A German chemist named Hermann Staudinger was the first person to publish theories on the molecular constitution of plastic materials in 1922. His theories centred on what he called the **macro-molecule** ,which are now called **polymers**. He maintained that the molecules of plastics are made up of many chains of carbon atoms.

Meaning

Plastic materials are either **thermoplastics**, which, like chocolate, can be heated to soften them and which set hard and can then be reheated to soften again and **thermosetting**, which, like an egg, once heated and then cooled cannot be reheated to make it soft again. The chemical structure of the two plastic materials is different. The thermoplastic material has chains of molecules which can slide past each other when heated and when cooled will stay in the new shape but will slide again if reheated. The thermosetting material has chains of molecules which are **cross-linked**. The chains are held together firmly. When heated these chains bond and they will not separate if reheated.

Associations

See: **curing, plastics, synthetic, thermoplastics, thermosetting, vulcanise.**

Cross section

Pronounced: KROS-SEK-SHUN (*o as in got, e as in let, u as in fun*)

Origin

Cross is from the Latin *crux*, which became *cros* in Old English meaning a *stake with two pieces of timber, one placed crosswise on the other*. Section is from the Latin *sectio* meaning *something cut off*. Compare with *bisect* and *disect*.

Meaning

A cross section is formed when an object (e.g. a piece of timber) is cut through horizontally or vertically at a right angle to the length or width of the object. One part of the object is removed, so that the inside of the other part can be clearly seen. In a drawing, the cut part is **hatched** with lines usually, but not always, at 45°.

It refers also to the view seen if an object (e.g. a building) is cut through its central axis at right angles to the viewer.

The term **sectioning** is more often used in mechanical drawing.

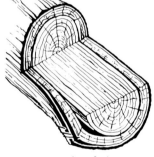

cross section of a log

Associations

See: **angle, drawing, extrusion, kerf, mechanical drawing, profile, projection, saw, technical drawing.**

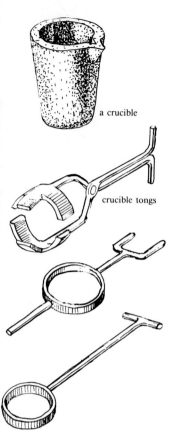

a crucible

crucible tongs

Crucible

Pronounced: KREW-SIBAL (*ew as in few, i as in lid, a as in ago*)

Origin

From the Latin *crux* meaning a *cross*. As a nightlight and melting pot used in Roman times was in the shape of a cross, eventually it was called a *crux* and then a *crucible*.

Meaning

A crucible is a receptacle for melting metals, often used in casting. It is made of a material, which is *refractory* (meaning resistant to heat), so that it can withstand high temperatures. It is made from cast iron, wrought steel or graphite.

Associations

A **crucible shank** is used for the pouring of a metal-filled crucible. **Crucible tongs** with an adjustable gripping and carrying device are used to carry crucibles filled with molten metal from a melting furnace to a pouring place.
See: **casting**.

Curing

Pronounced: KURING (*u as in pure*)

Origin

From the Latin *cura* meaning *care, concern*. The Latin *curare* meant *to take care of.* The word came to have the meaning of caring for something (e.g. crops, timber, leather etc.) so that it was not neglected but had its optimal use.

Meaning

Curing is a chemical reaction which produces a physical change by hardening or setting a substance (e.g. an adhesive). For example, a common synthetic resin, urea formaldehyde, requires a catalyst such as ammonium chloride to set or cure it. Curing can be produced by heat, by the addition of a catalyst or by uniting with oxygen from the air (e.g. oil-bound finishes, such as oil paints, varnishes and enamels, are hardened by exposure to the air - called **oxidation**). Curing is used for the hardening of concrete or plastic and also for a degree of vulcanization of rubber.

To cure timber is to ensure that it is dried correctly so that there is no warping.

Associations

See: **accelerator, adhesive, catalyst, cross-linking, distortion, laminate, moulding, oxidation, particleboard, plastic, resin, thermoplastic, thermosetting, warp**.

Current

Pronounced: KU-RUNT (*u's as in fun*)

Origin

From the Latin *currere* meaning *to run*. Compare with current meaning a running stream, and curriculum meaning the course of studies run in a school.

André Marie Ampè (1775 - 1836) a French mathematician and physicist, was the first to distinguish between current and potential difference. He discovered that a helix of wires carrying a current would act as though it was a bar magnet. He called the helix a **solenoid**. In 1800, Alessandro Volta, an Italian physicist, constructed the first effective electric battery and in 1827, Georg Ohm, a German physicist, defined the basic unit of electrical resistance.

Meaning

In electricity, a current is a flow of electrical energy from a source which starts and sustains the energy (called the **electro-motive force** - the e.m.f., which is measured in **volts**) along a conductor to a point which is lower in electrical energy than the source. The strength of an electrical current is measured in **amperes**.

A **direct current** flows in one direction. An **alternating current** reverses its direction rapidly at regular intervals, usually 100 times per second. A **rectifier** is a device which can change alternating current into direct current.

Associations

An **alternator** is an electrical generator which produces alternating current. A **galvanometer** is an instrument used for measuring small currents or differences in

voltage.
See: **circuit, electrolysis, fusibility.**

Cybernetics

Pronounced: SI-BUR-NET-IKS (*i as in silence, u as in fur, e as in let, i as in ink*)

Origin

From the Greek *kubernetes* meaning *steersman* - the one who controls the ship and says which direction it should take. The word was coined by an American mathematician, Norbert Wiener, in 1948.

Meaning

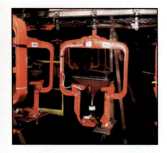

Cybernetics is a relatively-new branch of technology, developed from the comparison of the human nervous system with the operations of computers. It has no well-defined area of activity but draws upon mathematics, physiology, psychology, and electrical and computer engineering. It is a science of the systems of control and communication in animals and machines. The science's principal aim is to produce machines which are capable of functioning like a human. That is to be able to produce computers and robotic machines which are able to make decisions when presented with all the facts of a situation.

The science has produced machines which have allowed disabled people to control their own movements.

The development of integrated circuits and micro-computers has increased the development of robots and robotics will become an important part of technology in the near future.

Associations

See the work of the British cybertnetics scientist Meredith Tring (1915 -).
See: **machine.**

Dapping

Pronounced: DAPING (*a as in bat*)

Origin

From the word *dab* which comes from Middle English *dabben* meaning *to strike lightly*

Meaning

To dap is to tap and hammer lightly. Dapping is a process of shaping metal by lightly hammering the metal while it is in a shaping tool. For example, part-spherical shapes can be produced in metal by the use of a **dapping block** and hardened steel **dapping punches**. To dap is also the process of cutting notches in the construction of timber bridges.

Associations

A **swage** is a tool used to shape metals.
See: **burr, chasing, ductile, malleable, stake, swage.**

Design

Pronounced: DE-ZIN (*e as in delay, i as in sign*)

Origin

From the Latin *designare* meaning *to mark out*.
The term *design* was first used in the late 1600s, when there was a tendency for some people to specialise as the designers of products rather than the makers of a product. Before this time, a craftworker would both design and make a product.

Meaning

The word design has the following meanings:
- the intentional choice and arrangement of elements (e.g. line, shape, space, colour, texture, balance, etc.) in the composition of a drawing, photograph, or graphic presentation;
- the choice and use of materials to produce objects or create environments which meet human needs in functional and aesthetic ways (e.g. in, furniture, interior decoration, and industrial design);
- a project or scheme in which objectives and the means to attain the objectives are stated ("*marked out*");

- a preliminary outline (e.g. sketch or model) showing the main features of a project.

In brief, a designed product must work, be reliable, be good value for money and look good.

The design of a product usually involves the following problem-solving processes:

- A concept is formed;
- A design brief is formulated in which the problem, the constraints imposed and the objectives to be reached are made clear;
- Research is undertaken to obtain sufficient knowledge and understanding to consider several solutions to the problem;
- Analyis of the data to sift significant from insignificant information;
- Synthesis of data and and the consideration of several choices of action;
- Experimentation with possible solutions, using models, sketches, mock-ups etc.;
- Final design to meet the requirements as set out in a design brief.

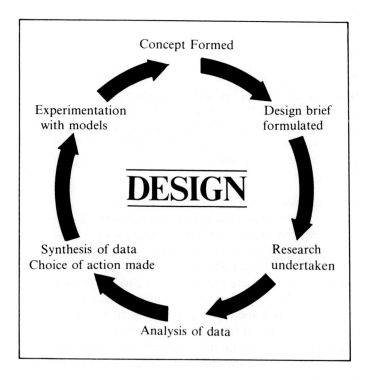

In Europe the word design tends to mean **industrial design**.

Associations

A **designer** is a person who through education and experience has the knowledge and understanding, techniques and skills successfully to complete design tasks for an agreed fee.

In Sweden, students are required to study **sloyd** (wood and metal design) from age 6 to 9. The aim of the system is to promote manual dexterity and awareness of good design in all students.

See the work of the following outstanding and influential, modern designers: **Industrial Design**: Peter Behrens (West Germany), Charles Eames (U.S.A.); **Interior Design**: Le Corbusier (France), Raymond McGrath (Australia); **Silversmithing Design**: Robert Welch (England), George Jensen (Denmark); **Furniture Design**: Alvar Aalto (Finland), Charles Rennie Mackintosh (Scotland).

See: **Bauhaus, ergonomics, etching, figure, fillet, Industrial Designer, inlay, layout, motif, pattern, relief, specification.**

Develop

Pronounced: DI-VEL-UP (*i as in pin, e as in help, u as in bonus*)

Origin

The origin of develop is uncertain. It may have originated with the Latin *dis* meaning *apart* and *volvere* meaning *to roll* or it may have come from the Italian *viluppare* meaning *to fold* or *roll up*. The word took on the meaning of unfolding in the sense of revealing.

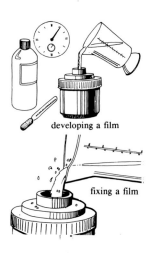

developing a film

fixing a film

Meaning

To develop is to bring something to an active or visible state. In photography it means to treat material to make pictures visible and to bring out the images. A **developer** is a chemical solution which is able to react with an emulsion on a film so that the latent (hidden) images in the emulsion are made clear when the film is exposed to light. Developing solutions vary according to the use to which they will be put but normally they contain 4

washing & drying the film

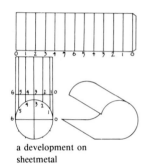
a development on sheetmetal

constituents: a **reagent** (a substance which causes other substances to react with each other), an **alkali** or **accelerator** (such as sodium carbonate), which speeds up reactions, a **preservative** to prevent premature oxidation (such as sodium metabisulphite) and a **restrainer**, which reduces the possibility of a developer making a film foggy. There is a golden rule in photography which says *exposure governs density and development governs contrast*. A **developing tank** is a container which holds film during a developing process. The tank allows developer to reach all of the film but it excludes light. In sheetmetal work, develop refers to the process of determining the shape of the sheet of metal to be cut to produce the design required. This is called "*developing the surface*". The finished drawing which is marked out on the surface of the sheetmetal is called the **development**. Developments are usually produced by using parallel or radial lines or by the method of triangulation.

Associations

Hydroquinone is a chemical used in developing which gives high contrast to photographs. **Reticulation** is the shattering of the images on a picture during or after the developing process, caused by a sudden drop in temperature. A **replenisher** is a solution which is put into a developer to bring it back to full strength. A **stop bath** is a weak acid solution in which a film is placed immediately after a developer has been used to stop the developer from working.

See: **accelerator, computer graphics, exposure, fixing, glaze, mechanical drawing, negative, notching, preservation, triangulation**

Die

Pronounced: DI (*i as in fine*)

Origin

From the Latin *datus* meaning *given* or *cast*, which changed in French to *de* and then in English to *die*, meaning *something cast in metal*.

Meaning

Die has a number of meanings, namely: a hollow mould used in casting metals into shapes required (often large numbers of identical objects); two segments of a hollow

a die

screw for cutting external threads of a screw or a bolt, either manually or by a machine; an engraved intaglio stamp which is used to make coins, medals or embossed paper; a metal form which determines the shape of plastics impressed on it or forced through it.

Associations

A **punch press** is an industrial machine for shaping metal using high pressure. A hydraulic or mechanical ram moves a punch against a die which is on the bed of the machine. This process is called **die casting**. **Blanking** is when a die is used to cut a shape from flat metal sheets or strips of metal. A **blank** is the shaped metal removed from the machine after stamping.

See: **blow moulding, casting, chasing, drawing, emboss, extrusion, impression, injection moulding, intaglio, moulding, notching, shape, stamp, swage.**

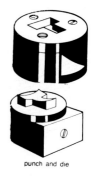

punch and die

Dip

Pronounced: DIP (*i as in lip*)

Origin

From Old English *dippan* meaning *to dip* or *immerse*

Meaning

To dip is to put or let down something into a liquid, e.g. a material dipped into a dye, paint or a liquid coating-substance, such as a plastic which is used for finishing. **Dip coating** is to put a surface of plastic on an object either by dipping the heated object into plastic powder, which fuses to the surface of the object, or by dipping a cold object (e.g. fabric gloves) into plastisol

(liquid plastic) and then using heat to cure the plastisol.
A **dip pot** is a pot holding a weak solution of flux into which a soldering bit is dipped, in order to remove oxide from the bit before soldering begins.

A **dip stick** is a scaled metal rod which is inserted into a tank or sump, in order to measure the level of liquid present.

Associations

See: **finish, flux, scale, solder.**

Distortion

Pronounced: DIS-TOR-SHUN (*i as in miss, o as in port, u as in fun*).

Origin

From the Latin *dis* meaning *in two ways* and *tort* meaning *twist*; that is *twisted in two ways*.

Meaning

Distortion is when a substance is made crooked, unshapely and put out of its natural shape.

If timber is dried too rapidly in a kiln, the outer layers of the timber are harder than the inner layers, which results in tensions within the wood which can produce distortion (e.g. **warping**) of the timber.

Distortion can occur in work in metal (or plastic) by over-heating, by too rapid cooling, where heating is not uniform but in one place (resulting in **upsetting**) or where an allowance has not been made for both lateral and longitudinal expansion after heating. **Controlled distortion** in welding is where metals to be joined are heated on the opposite side of the welded joint before welding takes place. The distortion that occurs from heating is balanced by the distortion caused by welding. In metal work **strength** refers to the ability of a material to resist distortion.

In electricity, distortion describes the jumbled and shrieking sound produced when a signal is amplified too much.

In photography, distortion refers to an image that is not prefectly formed by a lens. Distortion in photography can be deliberate, in order to produce a particular affect. The film can be deliberately processed incorrectly or the printing paper can be mis-shaped or mis-

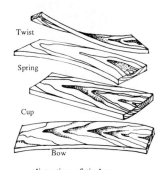

distortion of timber

aligned on purpose.

Associations
Cupping is distortion of wood during its drying, owing to unequal shrinkage throughout the wood. The wood *cups*; that is it changes to a slightly concave shape. Other distortions of wood are **springing, bowing, twisting** or **winding** and **splitting**.
See: **collapse, curing, durability, focus, foreshortening, grain, kiln, proportion, seasoning, shrinkage, warp, web, weld.**

Dowel

Pronounced: DOW-AL (*ow as in cow, a as in ago*)

Origin
From Middle English *dowle* meaning a *peg, pin* or *plug*. As the early carpenters had no metal nails or screws, they secured their wooden constructions with wooden dowels.

Meaning
A dowel is a headless pin or peg of wood or metal for

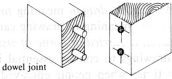
dowel joint

fastening together or keeping in position pieces of wood, stone, etc.. They are used to form an edge-joint of two boards (called a *dowel joint*).
Dowelling is cylindrical wood rods of about 6 to 25 mm.diameter, usually of birch or beech, which is sold in lengths of 3 to 4 metres.

Associations
A **dowel bit** is a short twist bit.
See: **bit, fastener, jig, joint.**

Draft

Pronounced: DRAFT (*a as in cast*)
Origin
Originally, the word was *draught* but the spelling has

been simplified over the years, although the word draught is still in use. The word is from the Middle English *draht* meaning *to draw, to pull* or *extract from*.

Meaning

A draft is a plan, sketch or preliminary drawing of a work which will be substantially developed and elaborated on and then produced. It also means the skill of producing a draft.

Draft also refers to the taper on the sides of a pattern which eases its withdrawal from a mould.

Associations

A **draughtsman** is a person who makes drawings, plans or sketches, often using a **drafting machine**. A draft in design is similar in many respects to a **study** in painting. See: **casting, clearance, drawing, mechanical drawing, moulding, pattern, sketch, taper, technical drawing**.

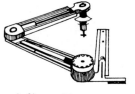

a drafting machine

Drawing

Pronounced: DRAW-ING (*aw as in saw*)

Origin

From the Old English *dragan* meaning *to draw, pull, carry or bear*. One meaning of drawing came to mean *what information is carried or borne in visual form*.

The use of drawings to convey essential information goes back to times when people engraved their ideas on walls of cave dwellings. Drawing, as a means of conveying and interpreting accurate and unambiguous data, has developed through the ages, until now it is an integral part of the language of designers, engineers and technicians.

Meaning

Drawing communicates ideas and feelings by giving us information in a visual form. It is very close to writing, especially in Chinese writing. Drawing or drafting (also spelled *draughting*) is the basis for most forms of visual expression. It is, however, a skill in its own right. Drawing is linear and is made more expressive by shading; painting, in contrast, is mainly the placing of masses of colour. Basically, to draw is to trace a line or figure by moving a pencil, pen or etching instrument across a surface in order to produce a picture or representation of an object. Drawing can also be produced by shading only. **Working drawings** are the complete set of drawings necessary to plan and manufacture an object or a machine. They should include: the title of the drawing, dimensions and instructions, the scale used, the projections used (e.g. Third Angle), the materials to be used, constructural methods (e.g. joints, fittings), the listing and shapes of component parts and sectional details, tolerances allowed, drawing sheet size, any special treatments needed (e.g. heating, painting) surface finishing required and also special data related to schedules, revisions and people involved in the various stages of the project. **Assembly drawings** (also called **general drawings**) show how the various components are to be assembled to form the completed work. Most working drawings are *drawn to scale*, usually being reduced so that large jobs can be fitted onto drawing paper. A full-size working drawing on plywood of a design (for example a piece of furniture) is called a **rod**.

Various **orthographic projections** are used but the most common nowadays is the **Third Angle Projection**.

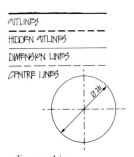

lines used in technical and mechanical drawing

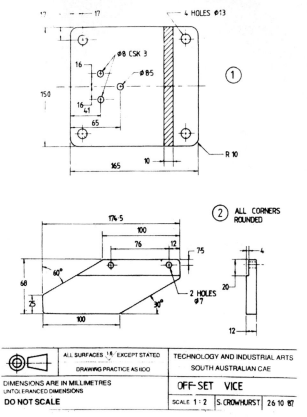

A **development** is a drawing on sheet metal and the process of making the total, unfolded drawing on the sheet of metal is called **developing the surface**. Complex drawings in a range of colours can now be produced by a computer-controlled **drafting plotter**.

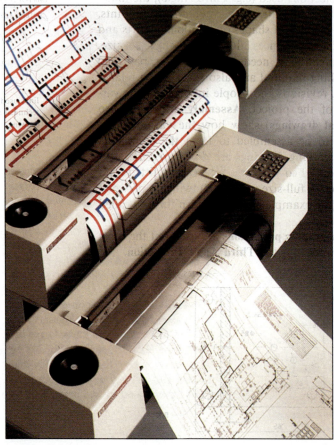

a drafting-plotter

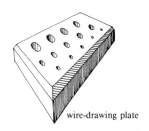

wire-drawing plate

In metalwork, drawing refers to the shaping and stretching of metal by heating and hammering it and to the tempering of metal. It refers also to the pulling of metal through dies with small holes to reduce the metal to wires and to the removal of a pattern from a mould.

Associations

An **extensometer** measures the elasticity of a metal; that is the extent to which it can be drawn.
See: **casting, cross-hatching, cross section, develop, die,**

draft, drawing, ductility, format, geometric, grid, hammer, Industrial Design, line, malleable, mechanical drawing, pantograph, pattern, perspective, plane, projection, rectilinear, repoussé, scale, shape, sketch, specification, temper, template, tensile, tolerance, trammel, triangulation.

Dress

Pronounced: DRES (*e as in best*)

Origin
From Middle English *dressen* meaning *to direct, prepare and guide.*

Meaning
To dress a material is to subject it to cleansing, trimming and smoothing. Stone, textiles, fabric, metal (cast metal in particular) and wood are *dressed* when their surfaces have been through a finishing process. Terms which are used to specify machining requirements are: *dressed all round, dressed on two faces, dressed on one face and one edge,* and *dressed on two faces and one edge.*

Dressed timber describes seasoned timber which has a smooth finished surface after having been planed by a planing machine.

The process of folding or bending the edges of metal with a mallet and stake is also called dressing, as is the process of *truing* the surface of a grinding wheel.

To dress a tool means to to return it to its original form and sharpness when new by forging or grinding it.

Associations

A **dresser** is an iron block used to forge work into shape on an anvil and a mallet for flattening sheet-lead.

See: **burr, face, finish, forge, grid, grind, hammer, panel, paring, plane, stake, tooling.**

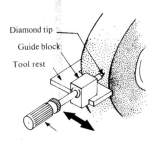

Dressing an abrasive wheel

Drill

Pronounced: DRILL (*i as in ink*)

Origin

From the Dutch *drillen* meaning *to bore*.

Simple hand drills were used thousands of years ago, being wooden shafts with points of sharp rock bound to them. The **bow drill** was used in Ancient Egypt in 2500 B.C. The **brace and bit**, which allowed constant rather than intermittent drilling, was invented in Northern Europe in the 15th century. Leonardo da Vinci is credited with inventing the first mechanical drill about 1495. The portable hand drill was invented by James Nasmyth (1808 - 1890) in 1844. The first electric hand drill was invented in Germany, by Wilhelm Fein in 1895.

Meaning

A drill is used for boring circular holes in substances. Nowadays, there are many kinds of drills, which have been produced specifically to bore holes in many different substances, from wood to granite. Drills today are made of high-speed steel or carbon steel, which contain heat-resistant tungsten, which allows them to operate at high cutting-speeds without the risk of burning. Drills in use in most workshops are: the straight-fluted drill (mainly used for drilling brass, sheetmetal and acrylic plastic), the twist drill, the centre drill and the countersink drill. A star drill, with a star-shaped point, is used to drill masonry and stone. It is usually driven with a hammer. Masonary drills, which have a special alloy head inserted as a cutting edge, should be operated in a power drill at a relatively slow speed to avoid overheating and possible damage to the drill.

STAR DRILL

using an impact drill with a masonry drill

Power drills are available in three categories: light, general and industrial. The drill size is measured by the capacity of the *chuck* and common sizes are 6mm, 8 mm, 10mm and 13mm.

The **drill press** is a machine found in most workshops. It is a geared, motor-driven power-tool with variable speeds, used mainly for drilling holes. It is fixed to a floor. The work is held in place on an adjustable worktable, attached to a fixed column of the machine. The drill is lowered onto the work.

The thickness of a drill is called the **web**. It is measured between the bottoms of the **flutes**.

using a drilling press

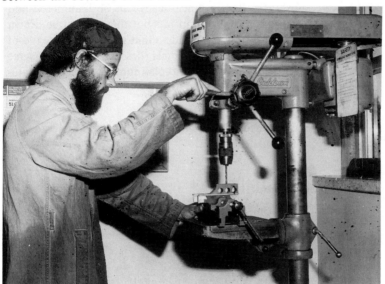

Associations

A **reamer** is a fluted cutting tool (available in a wide range of sizes and designs, e.g. parallel, tapered and expanding reamers), which is often used after drilling to produce a smooth accurate hole by removing small amounts of metal from the drilled hole. A hole is sometimes drilled to a size smaller than that required (e.g. approximately 0.025 cms from the size) and it is then enlarged with a reamer to produce a very accurate hole. A **core drill** is a hollow drill which removes a solid, cylindrical piece of metal (rather than metal-chips) from a piece of metal. It is used mainly for obtaining test specimens.

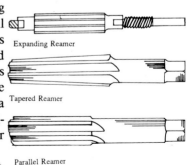

Expanding Reamer

Tapered Reamer

Parallel Reamer

See: **bit, bolt, boring, brace, chuck, counterbore, coolant, fixing, fluted, gauge, rake, stop.**

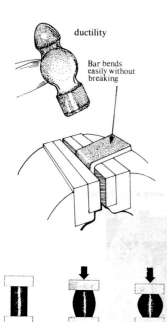

ductility

Bar bends easily without breaking

ductile materials tested under compression

Ductility

Pronounced: DUK-TILITY (*u as in luck, i's as in ink, y as in duty*)

Origin

From the Latin *ductilis* meaning *that which may be led or conducted; that which may be hammered.*

Meaning

Ductile describes the property of a metal that is capable of being drawn into a fine wire without breaking or fracturing. It is not the same as **malleable**, as some substances (for instance, lead) can be hammered and stretched into very thin bits but they cannot be drawn out into thin wire strands. Gold is the most ductile of metals. Other ductile metals (descending in the degree of ductility) are : silver, platinum, iron, copper, aluminium and nickel.

Associations

See: **extensometer** (in *drawing*), **gilding, hammer, malleable, tensile.**

Durability

Pronounced : DURA-BILITY (*u as in pure, a as in ago, i's as in ink, y as in duty*)

Origin

From the Latin *durus* meaning *hard* and the verb *durare* meaning *able to endure.*

Meaning

A substance which has durability is capable of lasting by resisting wear and decay. **Tungsten carbide** is the hardest man-made metal, being only slightly less hard than a diamond. Cutting tools are sometimes tipped with the material, as it cuts metals much faster than cutting tools made from high-speed steel. It is made under heat from carbon and **tungsten** (itself a very hard material, which is the most heat-resistant metal, melting at 3420°C (6200°F).

Durability describes the ability of timber to resist attack from insects (e.g. termites and borers), fungi and bacteria. Timber is classified into **durability classes**, according to the anticipated life of a timber when used externally and in contact with the ground. Durability also describes timber's ability to cope with heat, moisture, abrasion, bending, and distortion factors.

Associations

See: **distortion, hardening**

Eccentric

Pronounced: EK-SEN-TRIK (*e's as in pen, i as in ink*)

Origin

From the Greek *ekkentros* meaning *out of the centre*.

Meaning

Eccentric describes circles having different centres or a wheel with its axle not in the centre. Something out of centre is said to be **off set**.

An eccentric is a device on an engine which is used to change the rotary movement of a crank shaft into reciprocal movement on a slide valve.

Offset

Associations

Eccentricity is the deviation of the centres of two circles from each other.

See: **reciprocate**.

eccentricity
(out of centre)

Electrode

Pronounced: ILEK-TROD (*i as in pin, e as pen, o as in rode*)

Origin

From From the Greek *elektron* meaning *amber*, a yellow, translucent fossil resin. The discovery of electricity was made by the Ancient Greeks, who found that pieces of amber when rubbed together would pick up small objects. William Gilbert (1540 - 1603), an English doctor, discovered that sealing wax as well as sulphur had the same effect as amber. He called these substances *electrics. Electrode* is from the Greek *elektron* and *hodos* meaning *the way*. The term was coined by the famous English chemist and physicist Michael Faraday (1791 - 1867) to describe a conductor ("the way") in which an

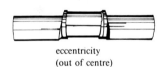

electric current enters or leaves an electrolyte, gas or vacuum.

Meaning

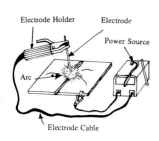

An electrode is the **conductor** (That is a device which can transmit or guide something.) through which electricity enters or leaves an electrolyte, gas, vacuum or other medium. It is either of the poles of a battery. A positive electrode is called an **anode**, and a negative electrode a **cathode**.

In welding, an electrode is that part of the electric welding circuit through which an electric current passes between the arc and the electrode holder.

Associations

An **electrolyte** is a substance that can dissolve to give a solution which is able to conduct an electric current. See: **anodise, circuit, conductor, electrolysis, electronics, grid, negative, positive, weld**

Electrolysis

Pronounced: ILEK-TROL-YSIS (*i as in pin, e as in pen, o as in pot, y as in duty, final i as in pin*)

Origin

From the Greek *elektron* meaning *amber* (See: **electrode**) and *lusis* meaning *a loosening, setting free, dissolving*. In 1807, the English chemist, Humphry Davy (1778 - 1829), became the first person to separate chemical compounds into their constituent elements by the use of electricity and thus prove the principle of electrolysis. The term was first used by the English Chemist and physicist Michael Faraday (1791 - 1867) on the advice of his friend, a Greek scholar, the Reverand William Whewell.

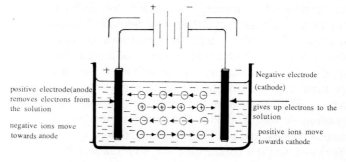

Meaning

Electrolysis is the separation (called *decomposition*) of materials contained in an **elecrolyte** solution by passing a current of electricity through it. All acids, bases and salts are electrolytes. Electrolysis is widely used in metal refining. For example, it is used to obtain aluminium from its ore, bauxite. It is also used in **electroplating**. Electrolysis is also called **electrolytic conduction**.

Associations

Electrolyte is a term commonly applied to the mixture of sulphuric acid and distilled water used in a car battery.
The decomposition of an electrolyte into positive and negative ions is called **ionization**.
See: **anodise, current, electroplating, electrode**.

Electronics

Pronounced: ILEK-TRONIKS (*i's as in pin, e as in pen, o as in pot*)

Origin

For the origin of *electron*, see **electrode**.
Electronics is the combination of *electron* and *ics*, which is a suffix used in Latin to form words explaining arts, sciences or branches of study, such as athletics, mathematics and politics.
The development of electronics owes much to the pioneering work in radio transmission and reception and the invention of carbon filament electric lights. Many scientists, such as as Heinrich Hertz (1857 - 1894), James Clerk Maxwell (1831 - 1879) and Robert Millikan (1868 - 1953) did much significant work, which led to the production of the first electronic devices. In about 1800, Thomas Edison discovered that carbon filaments emitted negatively charged particles when they were white hot. Almost 20 years elapsed before the particles were named **electrons**. Electronic devices, in the 1800s and early 1900s, were *gas-discharge tubes*. It was found that electric current could be made to flow in a vacuum tube. In 1897, the British physicist J.J. Thomson (1856 - 194) inferred the existence of the electron. Seven years later, John Ambrose Fleming made the first electronic valve or diode, and in 1906 de Forest

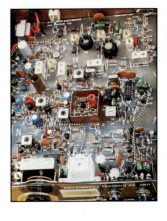

invented the triode. By 1922 manufacturers were producing millions of vacuum tubes a year for radio receivers. The invention of the cyclotron and the electronic microscope enabled great advances to be made in many scientific disciplines. In 1947, the transistor was invented by three Americans, W.H Brattain, John Bardeen and William Schockley (for which they received a Nobel Prize). It replaced vacuum tubes and made possible the vast development in integrated circuits and high-speed computers. Early in 1960 the first integrated circuits were produced.

Meaning

An **atom** contains a **nucleus** (comprising 99.97% of the mass of the atom), which contains positively-charged protons and uncharged neutrons. Electrons, which are negatively charged, are equal in number to protons and are outside the nucleus of an atom. An electron is an atomic particle, which has the smallest charge of negative electricity found in nature. It acts as the carrier of electricity in solids. Objects become **charged** by gaining or losing electrons.

Electronics is the branch of physics and technology which studies the movement of electrons in vacuums, gases, semi-conductors, circuits, etc. It is also the science of the application of electronic theory to practical devices such as radio, television, digital computing, x-ray machines, radar, automatic control systems and robotics.

Associations

The **electronic configuration** of an atom lists the number of electrons in each shell around a nucleus.
See: **circuit, computer graphics, conductor, electrode, negative.**

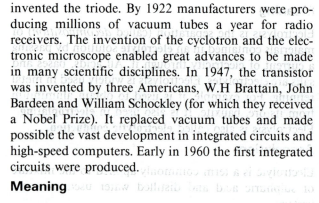

Diagram of a stable atom.

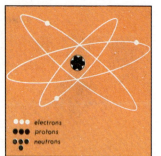

Electroplating

Pronounced: ILEKTRO-PLAT-ING (*i as in pin, e as in pen, o as in pot, a as in late*)
Origin
From the Greek *elektron* meaning *amber*, from which electricity was derived. (See: **Electrode**). *Plating* is from Old French *plate* meaning *a thin, flat sheet of metal.* In 1800 a German, Johann Wilhelm Ritter, discovered

76

the electroplating of copper. The English chemists, Humphry Davis in the early 1800s and Michael Faraday in the 1830s developed work which led to the use of electricity to coat one metal with another. Werner von Siemens, a Prussian officer, developed one of the first commercial electroplating processes in 1842. By the 1860s precious metals, such as silver and gold, were being electroplated onto cheaper alloys. In 1869 nickel plating began and chromium plating began in the mid-1920s.

Meaning

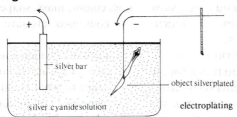

Electroplating is the most widely-used method of coating (or plating) one metal with another. **Electrolysis** is usually used. For example, to electroplate an object with copper, a pure copper plate and the object to be electroplated are immersed in a solution of a salt of the plating metal (i.e. a copper compound) which is an **electrolyte**. Electricity is passed between the copper plate and the object. The copper plate becomes the **anode** and the object the **cathode**. The electrolyte solution is decomposed and copper dissolves from the copper plate and is deposited gradually on the object, which becomes plated with pure copper.

Electroplating is used extensively, as it can protect metals from corrosion (e.g. a plating of chromium protects steel). It is also used in the making of tinplate for food cans, and in the making of jewellery. After plastic has been treated with acid to smooth its surface, it can be coated with a metal by electroplating. A.B.S. (acrylonitrile butadiene styrene) plastics have surfaces which are particularly suitable for metal deposits to adhere to them and very good finishes can be achieved with them.

Nickel plating is the only plating method safe enough and suitable for lessons in schools.

Associations

See: **anode, corrosion, electrode, electrolysis, finish.**

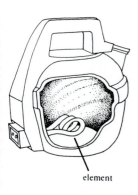

a heating element in an electric kettle. Elements (resistant wires) are used for heating purposes in kettles, irons, cookers, etc.

Elements

Pronounced: EL-I-MENTS (*e as in bell, i as in bit, e as in met*)

Origin

From the Latin *elmentum* meaning *a first principle* or *main constituent of something*, such as air, fire, water and earth.

Meaning

The elements of a visual composition are those basic visual qualities, such as line, colour, mass, shape, space and texture, which when combined produce visual images.

The term also refers to any substance which cannot be split up into a simpler substance. There are 92 chemical elements (classified into 9 groups in what is called the *periodic system*), which by themselves or in combination with other elements form every substance presently known.

Associations

See: **colour, composition, form, layout, line, mass, motif, pattern, proportion, shape, space, texture**.

Emboss

Pronounced: EM-BOS (*e as in lemon, o as on got*)

Origin

Boss is from Old French *boce* meaning *a swelling* or *bump*. Emboss means literally *to put bumps on something*.

Meaning

To emboss is any process (e.g. punching, hammering, moulding or using a die) which makes a design stand out in **relief** on metal, leather, paper, pottery, plastics, textiles or other material. Paper or other material can be embossed by placing the paper between a relief die and a hollow die. The relief die is struck into the hollow die to make a raised letter or design on the paper or material.
An **embossing** hammer is used for working on the inside surfaces of hollow, metal objects.

Associations

Bossing is a process of hollowing sheet metal, using a pear-shaped bossing mallet and a hollow block as a mould.
A **boss** is the centre or hub of a wheel and also a circular disc which can be attached to a mechanical part.
Bossing is the process of shaping malleable metals to conform to irregular surfaces. A **swage** is a tool used to shape metals.
See: **chasing, die, hammer, impression, malleable, moulding, relief, repoussé, shape, stamp, swage, symbol.**

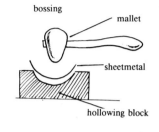

bossing
mallet
sheetmetal
hollowing block

plastic bosses.

Emulsion

Pronounced: I-MUL-SHAN (*i as in pin, u as in bonus, a as in ago*)

Origin

From the Latin *emulsio* from *emulgere* meaning *to milk out* or *to drain out*

Meaning

An emulsion is a milk-like mixture. It is the fine dispersion of one liquid in another (e.g. as in resin emulsion and latex emulsion paint) where the liquids do not mix with each other.
In photography, an emulsion is the light-sensitive coating on a photographic film or paper, which comprises a mixture of a silver compound (e.g. bromide, chloride, or iodide of silver) in a solution of gelatin. Photographic emulsions vary according to the subject being photographed, the lighting conditions and the purpose for which the photograph is needed. An **ordinary emul-**

sion is sensitive to blue, violet and ultra violet but not to red and yellow (for example emulsion on enlarging paper). **Orthochromatic emulsion** is sensitive to yellow as well as those blue colours in ordinary emulsion. Panchromatic emulsions (sensitive to all colours) are the most versatile and are most used nowadays.

Associations

Agitation in photography refers to the method of shaking a film tank to ensure that fresh processing solution is kept in contact with the emulsion on the surface of the film during a photographic process.

See: **bromide, exposure, filter, fixing, grain, hardening**.

Enamel

Pronounced: I-NAM-AL (*i as in ink, 1st a as in apple, 2nd a as in ago*)

Origin

From the Anglo-French *en* meaning *in* and *amel* or *email* meaning *to smelt or melt*. *Vitreous* is from the Latin *vitrum* meaning *glass*.

The origin of the process is not known. The earliest known enamels in the form of jewellery are from Greece and were made in the 13th century B.C.. The process was developed by Greek artisans in the 6th century and was then used extensively throughout the world. In 1799, Dr. Hinkling invented an enamelling process for saucepans.

Meaning

Vitreous enamel consists of a colourless, transparent compound called flux made of silica, potash or minium. The more silica there is in the compound, the harder it is. It is a very hard (vitreous) glass in powder form. It is coloured by metallic oxides, and the more oxide in the enamel the deeper the colour. Different oxides produce different colours: copper produces green, manganese produces purple, tin produces white and gold produces red. An object to be enamelled is cleaned with an abrasive (e.g. steel wool) and then covered with gum tragacanth, which ensures the enamel adheres to the object. The enamel is then fused onto the object by heating it from below with an air-acetylene torch or by firing it in a crucible in a kiln or a small, high-

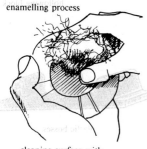

enamelling process

cleaning surface with steel wool

apply gum tragacanth with brush or spray

fine sieve enamel powder

temperature furnace called a **muffle furnace** at a temperature of about 200°F.. Usually enamels, which can be transparent, translucent or opaque, are glossy and brilliantly coloured. Enamel is used to decorate pottery, metal and glass and for making the preservative linings of objects. It is used extensively in jewellery work. It is an excellent insulating cover for electric wires.

Enamel is also pigments (to give colour) dispersed in resin, usually a natural or synthetic oily resin mixture. It is this vehicle which makes an enamel different from a paint. Varnish enamel dries by exposure to the air (*oxidation*) and the drying time can be decreased by the use of an accelerator. Lacquer enamel dries by evaporation and plastic emulsion by chemical reactions. The result is a smooth high-gloss film (although matt finishes can also be obtained) which is harder and tougher than the surface of paint.

heat from below at 850°C

Associations

If two colours of enamel are to be placed next to each other without a barrier, they may fuse during firing. To prevent this, a barrier of metal (called a **cloison**) is soldered between the enamels. This technique is called **cloisonné**. Another technique is to cut designs into a metal ground and then fill the incisions in the metal with enamels. This technique is called **champlevé**. Vitreous enamel objects are also decorated by using **stencils** and also by a **graffito** or **sgraffito** technique, where designs are scratched onto an object.

an enamelling kiln

Enamelling became popular during the Art Nouveau period in the early part of this century in Europe, particularly in France, where the work of René Lalique (1860-1945) was outstanding. One of the most famous **enamelists** in the world was Peter Carl Fabergé (1846-1920) who was born in Russia. He worked in precious metals and is best known for his superbly-crafted Easter eggs. See also the work of Alexander Fisher (1864-1963), a British enameller.

cloisonné work.

See: **accelerator, cladding, etch, filigree, finish, flux, kiln, lacquer, oxidation, pickle, pyrometry, solder, stencil, vehicle.**

Engraving

Pronounced: EN-GRAV-ING (*e as in pen, a as in brave*)

using an engraver

Origin

From the French *engraver* meaning *to carve* or *to cut*. Also, it is from the Old English *grafan* meaning *to dig*. Think of a *grave* for burial or a *groove* in a record.
The earliest known engraving is of "*Christ Crowned with Thorns*", dated about 1446. The Italian artist Parmigianino developed the technique from 1520 onwards. The earliest known copper engraving in England was made in 1461. The art of hatching and cross-hatching was invented by Michael Wohlgemuth of Germany in 1486. The technique was used in Europe in the fourteenth century to decorate armour.

Meaning

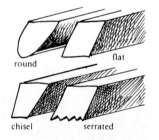
round   flat
chisel   serrated

Engraving is a technique in metalwork using a tool called a **graver** or **burin**, which is a chisel-like tool with a rounded wood handle and a blade sharpened to 45°, with a round, flat, chisel or serrated edge. An engraver cuts a composition, design or lettering into metal, wood or stone. Engraving is most successful on copper and zinc because they are relatively soft metals and easy to work, but they wear away fairly quickly, so they are sometimes faced with steel in order to lengthen their lives. Very fine hatching and cross-hatching can be made on metals to produce shading effects.

Associations

Xylography is the art of wood engraving; **chalcography** is the art of engraving on copper or brass. The three types of engraving are **intaglio**, **relief** or **cameo** and **surface** or **planar**. A **riffler** is a special file used in engraving. It is slightly curved and has narrow points at each end. It is especially useful for getting into difficult corners, for enlarging holes and for the filing of inside surfaces.
See the engravings of William Hogarth (English 1697-1764) and Albrecht Dürer (German 1471-1528).
See: **burin, chisel, composition, cross hatching, half tone, impression, laser, resist, scoring, serration, wood block.**

a riffler

Ergonomics

Pronounced: URGO-NOM-IKS (*u as in fur, 1st o as in go, 2nd o as in got, i as in ink*)

Origin

From the Greek *ergon* meaning *work* and *oikonomos* meaning *one who manages* (e.g. a household). Literally, it is how work is arranged and managed.

Ergonomics as a science began in World War 2 (1939 - 1945), when problems related to the mismatch between machines and their operators became apparent and had to be solved. Work done in ergonomics was later related to military and aerospace matters, but it is now related to most industries and commercial organisations, which are concerned that workers should operate most effectively in specific environments. In the United States ergonomics is called **human factors engineering**.

Meaning

Ergonomics is the study of the relationship between individuals and their work, their working environment and the tools or equipment (machines and furniture) they use in their work. It is particularly concerned with the problems which people experience in their work from physical factors (e.g. the construction of benches or tables at which people must stand or sit to work), mental factors (e.g. the monotony of conveyor-belt work) and psychological factors (e.g. noise, colour of surroundings, ventilation and light). A work bench, for instance, can be said to be *ergonomically sound* if its form is appropriate for the activity and is such that it allows someone to work at it for relatively long periods without the person experiencing physical strain. A chair is ergonomically well-designed if it can be adjusted to respond to changes in sitting position and gives support to the body at all times and can be adjusted for "*popliteal*" height (the distance from floor to thigh in a seated position when the feet are comfortably flat on the floor and the thighs and forearms are approximately horizontal); if it has an anatomically-formed seat and back to maintain a correct spine curvature, especially in the lumbar region and has no sharp corners of protruding parts which might be dangerous.

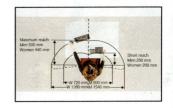

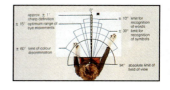

Software ergonomics deals with the ease with which a software computer package can be used by a computer operator.

The study of ergonomics is essential for designers who produce products which people will use.

Associations

Biotechnology is the application of technology to living things, including humans. **Bio-engineering** is similar. **Anthropometrics** is the science of the measurement of the human body.
See the work of Henry Dreyfuss (American 1903-1972).
See: **computer graphics, design, Industrial Design, proportion.**

Etching

Pronounced: ET-CHING (*e as in pet, i as in ink*)

Origin

From the Dutch *etzen* or the German *atzen* meaning *to eat* or *to bite*. European metalworkers began to use etching extensively for the decoration of armour and weapons in the 15th century.

Meaning

To etch is to produce a design on metal by the use of a corrosive acid (e.g. nitric acid). The metal surface must be polished before the etching begins, as polishing reduces the effectiveness of the work after the design has been completed. A composition or design is scratched onto metal which has been treated with an acid-resistant substance, such as beeswax, varnish or asphaltum. Where the resist substance is scratched, the metal plate is exposed. When the metal plate is dipped in acid, the acid "bites" only into the exposed lines.

Acid etching is also part of part of surface preparation of some metals before finishing. As well as cleaning a metal, the application of an acid (e.g. phosphoric acid) to a metal surface slightly roughens the surface to provide a "tooth" for paint to adhere firmly. Etching is particularly necessary for cold-rolled steel, tinplate, galvanised iron and aluminium because their surfaces are so smooth that they do not provide sufficient anchorage for finishers. Acid etchers (usually phosphoric acid or metallic phosphates) can be used in the form of an **etch primer**, so two functions can be performed in one operation.

etched metal jewellery

Associations

An **etcher** is a person skilled in etching. An etching process can also be used in enamelling and jewellery work.
See: **burin, composition, cross hatching, design, enamel, finish, gilding, intaglio, mordant, resist**.

Exposure

Pronounced: EX-PO-ZHA (*e as in let, o as in go, zh as s in fusion, a as in ago*)

Origin

From the French *exposer* meaning *to show, display, reveal*

Meaning

Exposure in photography is subjecting a light-sensitive photographic film, plate or paper to light. The exposure is the amount of light which is allowed by the camera (through a lens) to react with the emulsion on the film. A **shutter** is a device which controls the length of time light is allowed to reach the film. The shutter mechanism is operated by a **shutter release**. Exposure is measured by time and an **exposure meter** can be used to calculate fairly accurately an exposure needed. **Over exposure** results in a film receiving too much light and images which are light where they should be dark and vice versa. This is termed **solarisation**. A photographer can control the amount of light which parts of a film receive. When one part is allowed more light than others so that it becomes darker, the term used is **printing up** or **burning in**.
Nowadays, most cameras have automatic focusing and exposure systems which are controlled by electronic circuitry.

Associations

A **safelight** is light used in photography which will not affect photographic materials. A **step wedge** is a series of exposures on a strip of film or photographic paper, increasing by steps in the duration of exposure as a test piece to find the correct exposure needed.
See: **bromide, contrast, develop, emulsion, gradation, lens, stop**.

an over-exposed film

a light meter.

Extrusion

Pronounced: EX-TROO-SHUN (*e as in net, oo as in soon, u as in fun*)

Origin

From the Latin *extrusus* from the verb *extrudere* meaning *to push out, to protrude, to thrust out, drive away*

Meaning

An extrusion is anything which is **extruded** or forced out of something or somewhere. Extrusion in metal or plastic work is a process where hot or cold plastics or metals are forced through a die or other kind of moulding to produce required configurations and patterns, such as rods, tubes, pipes and various solid and hollow sections.

An **extruder** is a machine for producing a continuous shape of uniform cross-section.

extruded plastic pipes.

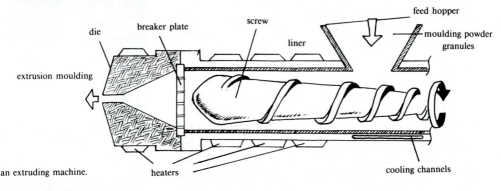
an extruding machine.

Associations

Pultrusion is a method of extruding plastic material in a continous stream.

See: **blow moulding, cross section, die, injection moulding, moulding, plastic, thermoplastic.**

Fabric

Pronounced: FAB-RIK (*a as in fat, i as in ink*)

Origin

From the Latin *fabrica* meaning *a workshop* and *faber* meaning *a worker in hard materials*. Eventually, the word came to mean anything made by skill and labour, and then cloth which was made by hand through weaving.

Meaning

Fabric is cloth material (textile) made by weaving raw materials, such as wool, cotton, flax, silk, hemp or jute. Fibre which is spun on a spinning wheel is called **yarn**. Fabric refers also to anything which is made or constructed by art and labour, for example the frame and structure of a building.

To fabricate means to construct or manufacture (which originally meant *to make by hand*). It also means to build up an object by joining parts together.

Associations

See: **joint, warp, wood block.**

Face

Pronounced: FAS (*a as in make*)

Origin

From the Latin *facies* meaning *a making, a shaping*. It came from the verb *facere* meaning *to make* or *to do*. The *face* side came to mean the side from which "*shaping*" began.

Meaning

To face a material is to give it a flat *surface* by machining it.

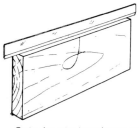
Testing face edge for straightness with straight edge.

The *face* of a tool or implement is its working surface and invariably its distinctive surface, for example, the striking surface of a hammer and the sole of a plane. The face of a metal casting is the surface which is turned or polished. The face of a gear wheel shows the width of the geer teeth.

wood with a face mark

A **face side** of a piece of wood is the side selected for its best qualities. All measurements for the piece of wood are taken from the **face side**. It is usually marked with a loop called a **face mark**. A **face edge** is a true edge in relation to the **face side**. It is usually marked with a line continuing from the loop marking the **face side**.

Associations

Facing is a lathe technique where ends of work pieces are finished by making them smooth.

See: **dress, facet, finish, gauge, lathe, machine, moulding, polyhedron, turn, veneer.**

Facet

Pronounced: FAS-ET (*a as in mass, e as in let*)

Origin

From the French *facette* meaning *a little face*.

Meaning

A facet is one of a number of flat surfaces (faces) on a gem stone or a crystal. Precious stones (e.g. diamonds) are cut so that they have facets, which reflect and refract (i.e. break the direction of) light to make the stones sparkle. Cutting tools have facets, which are bevelled edges. One facet can be used for grinding; another, at a different angle, can be used for sharpening purposes.

Associations

See: **angle, bevel, face, grind, spectrum.**

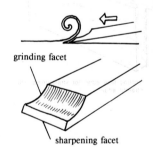

grinding facet

sharpening facet

Fasteners

Pronounced: FAS-NAS (*1st a as in last, 2nd a as in ago*)

Origin

From Old English *faestnian* meaning *to fasten, to fix*

Meaning

To fasten is to tie or fix by some bonding arrangement, such as adhesion, cohesion or mechanical linkage. Fasteners are used to bond wood, metal and plastic. The

fitting connectors

common fasteners readily available are wood screws, metal, machine screws, nuts and bolts and rivets. Nowadays, there is a wide range of **connectors**, which are assembly aids for the making of *knock-down* furniture units, which people can assemble themselves from purchased kits of parts.

An important form of fastening in many machines is the **clevis**. A **shackle** is a fastening device which allows some degree of movement.

Associations

See: **bolt, brace, cramps, dowel, fit, furniture, hasp, hinge, nails, plug, rivets, screw, tap.**

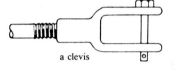
a clevis

Fatigue

Pronounced: FA-TEEG (*a as in cat, ee as in see*)

Origin

From the Latin *fatigare* meaning *to weary, tire, vex, worry*. *Fatigos* meant *driving towards exhaustion*.

The study of stresses in metals goes back to the 16th. and 17th centuries but it was towards the end of the 18th. century and then increasingly in the 19th. century with Industrialisation that research work was undertaken into stress in metals and into metal fatigue. The understanding of metals has increased enormously during the 20th.century with the discovery of X-ray diffraction, which was applied to the micro-structure of crystalline solids by an Australian scientist, Professor L.Bragg. The invention of the electron microscope has greatly widened the study of metals and increased understanding of fatigue problems

fractured metal owing to fatigue

Meaning

Fatigue describes the tendency for a metal to weaken and to crack or fracture when subjected to repeated or varying stresses caused by a cycle of changing loads. The **fatigue limit** of a metal is the end point of a range of stress which a metal can withstand indefinitely before it breaks down.

Associations

Tests for metal fatigue usually use methods of bending metal and vibrating it until the endurance level of the metal is reached. Nowadays, electro-magnetic and hydraulical methods of testing are common.

See the pioneering work of Navier (1785 - 1836), Cauchy (1789 - 1857), W.J.M.Rankine (1820 - 1872), J.C.Maxwell (1831 - 1879), Mohr (1835 - 1918). See the Wöhler test, which uses a ball-bearing and the Haig fatigue testing machine, which uses a powerful elctromagnet.

See: **metallurgy, tensile.**

Ferrule

Pronounced: FE-ROOL (*e as in pen, oo as in soon*)

Origin

From the Latin *ferrum* meaning *iron* and *variae* meaning *a little bracelet*.

Meaning

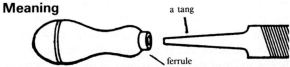

A ferrule is a metal band or ring, usually of steel, which is used to strengthen tools with wooden handles (e.g. chisels, files, paint brushes, screwdrivers and soldering bits) at the joining of the wooden handle to the metal part. At this point the wood or metal handle is usually tapered. It is called the **tang** and could be a weak point in the tool without some support.

Associations

See: **fit, taper.**

Figure

Pronounced: FIG-A (*i as in big, a as in ago*)

Origin

From the Latin *figura* meaning *form, shape*.

Meaning

A figure is any external form or shape in outline. It is a two-dimensional space enclosed by lines or a three-dimensional space enclosed by surfaces (e.g. a pyramid or sphere). It also refers to a diagram, an illustrative drawing or a decorative pattern.

In woodwork, it describes the natural markings on timber; the designs produced in finished wood by growth rings, the grain and texture and colour.

Figured means embellished with a pattern, as swords and armour were in former times.

Associations

See: **design, grain, motif, pattern, shape, veneer.**

Filament

Pronounced: FILA-MANT (*i as in fill, a's as in ago*)

Origin

From the Latin *filamentum* meaning *a thread-like body*, which comes from *filare* meaning *to spin*.

Early in the 19th century Sir Humphry Davy invented a carbon arc light. A light from the spark (or arc) in the passage of an electric current passed between two rods of carbon By 1850, an English Chemist, Joseph Swan, had invented, with limited success, a filament incandescent lamp. In the U.S.A. in 1878, Thomas Edison produced a filament made from a strip of carbonised bamboo, which produced an effective and relatively long-lasting light.

Meaning

A filament is any slender, thread-like material, such as a single fibre in the structure of a plant.

In electricity, a filament is a very fine wire of high resistance which is not easily fusible. It can be heated until it glows (is *incandescent*) by the passage of an electric current. An incandescent light operates on the

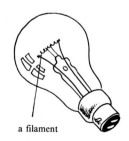

a filament

principle that when a substance resists the passage of an electric current heat is generated. If the substance is heated until it becomes white hot, it glows or incandesces. In many electric lights, a filament is the source of the light, as in an electric globe or bulb. It is also the means by which electrons are emitted from a thermionic valve or tube to generate heat, as in an electric fire. Many filaments today are made from tungsten, a metal which takes a high temperature.

In plastics, a monofilament is a plastic material which has been forced through a small hole to form a single fibre.

Associations

See: **current, electron.**

Filigree

Pronounced: FILI-GREE (*i's as in sit, ee as in see*)

Origin

From the Latin *filum* meaning *a thread* and *granum* meaning *a grain*. Filigree is an abbreviation of *filigreen* or *filigraine*. Originally, it meant delicate work made in threads and beads or grains. The process was used in Ancient Greece and Rome. It was a popular form of ornamentation for jewellery and vessels among the Anglo-Saxons in the 7th and 8th centuries. Venetian metalworkers in the 13th and 14th centuries produced filigree which was enriched by enamelled cloisonné work.

William Fountain, Great Britain, *Dessert dish cover* 1803, sterling silver, Art Gallery of South Australia, Adelaide, South Australia

Meaning

Filigree is the very fine and delicate use of elaborate openwork patterns (with lace-like effect) using metals which are sometimes precious. Such work is usually produced from fine, flattened, twisted wires which are made into patterns and then soldered. The work is either attached to a solid backing or left as unsupported open work. It is used extensively in jewellery-making and in decorative metal work for gates, fences and varandahs. **False filigree** is produced by the **lost wax** casting process, where delicate forms are modelled in wax.

filigree work

Associations

See: **cire perdu, enamel, pattern, solder**.

Filler

Pronounced: FILA (*i as in pin, a as in ago*)

Origin

From the Old English *fyllan* meaning *to fill*.

Meaning

A grain filler is a material used to fill holes or pores in a timber surface (especially open-pore woods such as oak or walnut), so that it has a level base on which to apply a finishing substance. A filler substance should not react chemically with the finishing material which will be used, so the manufacturer's instructions on a finishing product should be carefully observed. The two types of filler most used are **linseed oil powder-paste fillers** and **plastic-based paste fillers** (which are used under plastic finishes). A filler can be coloured to match a wood stain, so that a colour change can be made to a wood to enhance its beauty. A filler is also a substance to increase the bulk of something, usually so that a better fitting can be made. The term refers also to a material added to paint to make it thicker and to gravel cinders, broken bricks and rocks used as a base for concreting.

Fillers are also solid materials which are added to resins or polymers for specific reasons, such as to give them electrical or heat resistance (e.g. mica and asbestos), to strengthen them (e.g. fibre-glass), to make them more

flexible (e.g. rubber) or to make them self-lubricating (e.g. graphite).

Associations

Beaumontage is a hard filler (made of beeswax, melted resin and a little shellac) for nail holes or other marks in a wood surface. **Silex** is a form of silica used widely in making linseed oil paste wood fillers. Filler material is usually added to the surfaces of **hardboards** and **fibreboards** to make them less porous (which saves paint) and also to make them easier to paint. An **extender** is an additive which increases the bulk of a substance without altering any of its properties. A **filler rod** is metal wire which is melted into a weld to strengthen the weld by increasing its bulk.

See: **brazing, finish, joint, paint, pigment, porosity, resin, size, weld.**

Fillet

Pronounced: FILIT (*i's as in ink*)

Origin

From the Latin *filum* meaning *a thread* or *thread-shaped*. That is something long and slender.

Meaning

A fillet is a raised and partially-rounded rim or ridge (concave or convex) between two intersecting surfaces. It is a structure added to round off the interior angle of a joint, which gives added strength and an attractive design.

It refers also to a piece of timber which is attached to work to support a shelf, and to a narrow, flat or square projection or band on a moulding. Also, it is a small strip of wood which covers the end grain in a joint.

A **fillet weld** is an almost triangular-shaped weld that joins two intersecting metal surfaces making a lap.

Fillets are made to give added strength to metal joints (e.g. soldered joints) and plastic joints, so that they can withstand stresses.

Associations

See: **angle, design, joint, lap, moulding, solder, weld.**

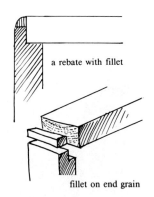

a rebate with fillet

fillet on end grain

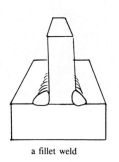

a fillet weld

Filter

Pronounced: Filta (*i as in pin, a as in ago*)

Origin

From Old French *filtre* meaning a *strainer made from felt*, which is a cloth made without weaving where fibres of a fabric cling together.

Meaning

A filter is a device for removing dirt, grit and unwanted bits of a substance from liquid or gas by passing them through a porous material, such as unsized paper, sand or charcoal. The liquid or gas is said to be **filtrated**.

A filter in photography is a coloured, transparent screen used to modify the colour of light which falls on a sensitised plate or film. Filters usually comprise a sheet of dyed gelatin which may be placed between two pieces of glass. Liquid filters are sometimes used in scientific work. The filters may be placed in front or behind the camera lens (usually in front) , directly over the plate or film or inside an enlarger. A filter of similar colour to the image lightens the shade and reduces contrast; a filter of a complementary colour darkens the image and increases contrast. In colour photography, the three-colour separation negatives are made through tri-colour filters (red, green and blue), each of which transmits one third of the spectrum.

Associations

A **polarising filter** is a piece of glass which is treated with a thin film of a substance which cuts out glare reflected from non-metallic shiny surfaces.

See: **bromide, colour, contrast, emulsion, grid, lens, negative, porosity, spectrum.**

Finish

Pronounced: FINISH (*i's as in ink*)

Origin

From the Latin *finis* meaning *end* or *finish* from the verb *finire* meaning *to finish, fix* or *end*.

Finishing processes were used in Ancient Egypt, Greece, Rome and Mesopotamia (using pitch, wax, asphalt, paints, varnishes and metal sheaths) not only for appearance's sake but to preserve precious objects and tools. Technological developments in chemistry and electricity in the 19th and 20th centuries led to the production of enamels, lacquers, synthetic resins, emulsion paints and to electro-chemical surfacing processes.

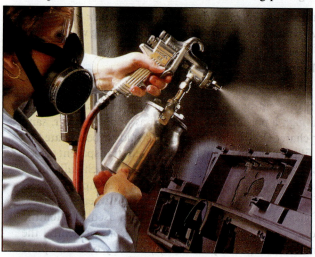

In electrostatic spraying, a part is first coated with a **conductive** primer. The part is **grounded** and an **electrically-charged** coat of paint is applied. The electrical attraction of the paint to the part eliminates up to 70% of overspray, which occurs in many conventional spraying techniques.

Meaning

To finish a piece of work is to put the final touches to it, first by ensuring all surfaces are clean, true and smooth (using cutting tools, finely-graded abrasive materials and sometimes chemicals such as a pickling acid) and then usually by polishing or by applying by brushing, spraying or dipping the last coat of plaster, enamel, glaze, paint ,dye or other finisher. Careful surface preparation is essential if finishing is to be effective. Some metals and plastics are finished by anodising and by using electro-plating. Finishers are used for decorative purposes (e.g. etching, knurling, engraving, enamelling, embossing) but also to preserve materials and to ensure they are clean and hygienic. The type of finish used depends upon a range of factors, including the material to be finished, the location of the finished object (external or internal), the purpose of the object, its intended life-span and the desired appearance (e.g. high gloss, matt, etc.).Wood finishes are applied to seal the pores in wood and minimise movement in the wood,(to *stabilise* it) to protect a surface from scratches and marks, and to enhance the natural beauty of wood. Some finishing materials for wood are: stains (colour stains, water stains, oil stains, spirit stains), beeswax (often with turpentine), oils (e.g. linseed, teak), cellulose

lacquers, varnish, shellac and synthetic resin polishes. Traditionally, finishes were rubbed into the pores of wood over a long period (e.g. in **French polishing**). Nowadays, most wood finishes are surface finishes.
Metals can be finished by applying **finishing cuts** with cutting tools or machines and by **burnishing**, **buffing**, **sand blasting** and **lapping**. Also special finishes, such as *hammer*, or *wrinkle* or *crackle* finishes can be applied. A **finisher** is a person or machine doing the last operation in a manufacturing process.

Associations

See: **abrasive, anodise, annealing, buffing, burnish, dress, electroplating, enamel, etching, face, filler, lacquer, laminate, lap, paint, paring, pickle, pigment, plane, porosity, preservation, priming, resin, sealer, size, stop, synthetic, varnish, vehicle.**

Fit

Pronounced: FIT (*i as in bit*)

Origin

From Old Norse *fitja* meaning *to knit together*, related to Old Dutch *fitten* meaning *to suit*.

Meaning

To fit is to be shaped in such a way and to be of such a size that one thing fills up a space in another, *mating* things exactly. There is said to be a fit when one thing suits another and corresponds perfectly with it, as when the threads on a bolt mate with those on a nut.

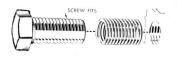

In metalwork and woodwork, fit has a specific meaning, as there are several kinds of fits. When parts of a work are put together the degree of tightness between the parts depends upon the function of the parts. Some mechanical functions can tolerate a **clearance** between parts; others cannot. For example, it would be acceptable for a shaft rotating in a bearing to have a looser fit than that of a tool blade into its handle. There are two categories of fits. First, an **Interference** fit is where a fit is obtained by forcing parts together (e.g. a **force fit** and a **drive fit**) or by shrinking one part on to another (e.g. a **shrink fit**) or expanding one part on to another (e.g. an **expansion fit**). Second, a **Clearance fit** is where there is a *clearance* or an *allowance*, which

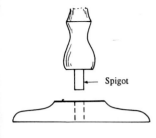
Spigot

allow parts to run relatively freely of each other (a **free fit**), or slide into each other (a **medium fit**) or where parts can be made to fit by slight force, such as by hand (a **snug fit** or a **push fit**). A **running fit** or a **sliding fit** describes parts which fit with just enough clearance to allow motion.

Associations

Play refers to the unplanned movement between poorly-fitted or worn parts. **End play** refers to the clearance allowed at the end of revolving parts, in a longitudinal direction. **Tolerance** describes the amount of variation allowed in the size or the clearance of parts.

A **shim** is a thin piece of metal or other material which is inserted between two parts to produce a tighter fit. It can also be used under an object to make it level.

See: **bit, clearance, fasteners, fixing, honing, joint, machine, mandrel, shrinkage, taper, tolerance.**

Fixing

Pronounced: FIX-ING (*i as in ink*)

Origin

From Latin *fixus* meaning *fastened.*

Meaning

Fixings are any devices which make anything firm, fast and stable. They are used to secure metal to wood, to other metal and to plastic, leather and fabrics, either permanently or temporarily. In the past there were three main groups of fixers or fasteners: **bolts**, **screws** and **rivets**.

Nowadays, there is also a wide range of metal and plastic **connectors** available. Fixings are frequently used in combination with some type of **plug**.

In photography, fixing is a process after developing of treating photographic film or paper with a **fixer**, so that the film or paper will not be chemically affected by light. A common fixer is **hypo** (sodium thiosulphate), which washes away unused emulsion and makes a picture permanent.

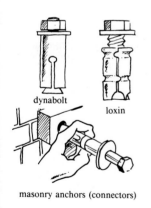
dynabolt
loxin
masonry anchors (connectors)

Associations

See: **bit, bolt, bromide, cleat, cramps, develop, drill, emulsion, fasteners, fit, hasp, nails, pinning, plug, rivet, screw.**

Flange

Pronounced: FLANJ (*a as in pan*)

Origin

From the Old French *flanc* or *flanche* meaning *the flank or side*

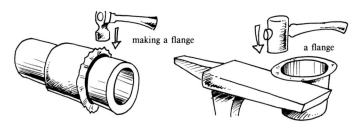

making a flange

a flange

Meaning

A flange is a flat rim or raised edge which stands out from the main body of a workpiece, such as a rim on a wheel which runs on a rail, the disc-shaped rims on the ends of pipes, so that they can be joined and a nut which a has a flange as a built-in washer. A flange is used for strengthening an object or for attaching something.

A **flanged joint** is any joint made between pipes which have flanged ends and are bolted together.

Associations

See: **burr, joint**.

Flitch

Pronounced: FLICH (*i as in ink*)

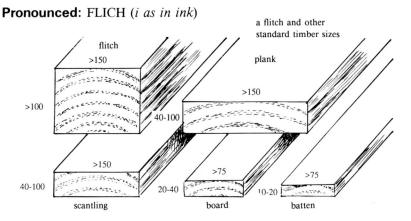

a flitch and other standard timber sizes

Origin

From Old English *flicce* meaning *the side of a hog which is salted and cured*. As the rough slab of meat resembled a slab of rough timber, the term was applied later to a piece of timber with its bark on.

Meaning

A flitch is a large, rough section of timber cut from a tree. It has at least two sawn surfaces. It is sometimes called a **slab** or a **junk**.

It is a term used commercially for one of the standard sizes of timber, which are, in decreasing order of size: flitch, plank, scantling (or **baulk** when a heavy scantling), board, batten and strip.

Associations

See **scantling**.

Flute

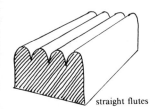

straight flutes

Pronounced: FLOOT (*oo as in moon*)

Origin

From the Middle English *floute* meaning a *flute*. As the the musical instrument the flute was long and thin, the word was probably applied to thin grooves of similar shape.

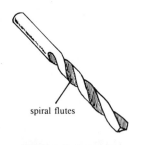

spiral flutes

Meaning

A **flute** is a semicylindrical, vertical groove in a pillar or in wood. **To flute** is to form curved furrows in a material. **Fluted** is ornamented with channels or grooves. Wooden and glass artefacts are often fluted for decorative purposes. Flutes are sometimes machined (with either a straight or a helical groove) into a cutting tool (e.g. a drill or tap) in order to make the removal of metal chips easier to escape and also to act as a channel to allow cutting fluid to reach the cutting point of a tool.

A **fluting machine** is a machine for corrugating sheet-metal and also a wood turning machine for forming twisted spiral and fluted balusters (a small pillar used as a support to the rail of a staircase).

a fluted column

Associations

See: **drill, groove, machine, mill.**

Flux

Pronounced: FLUX (*u as in dug*)

Origin

From the Latin *fluxus* meaning *flowing*.

Meaning

The most common fluxes are calcium, potassium, sodium, lead and boron. When a metal is heated (by annealing, soldering or welding), an oxide of the metal forms on its surface (sometimes referred to as *firescale*). It is rather like rust. The oxide prevents metals from fusing in a heating process. Fluxes (which are often a borax mixture in paste or liquid form) which clean the oxide from the metals, so that they can be joined together by solder, are called **corrosive fluxes** (e.g. hydrochloric acid, zinc chloride and ammonium chloride, which is also called sal ammoniac). Some fluxes, however, do not remove oxides before soldering but only stop them from forming during the soldering operation. These are **non-corrosive fluxes** (e.g. resin).

Associations

Entetic is a word to describe a substance which when mixed with another substance will melt and fuse at a lower melting point than will the separate substances.
Borax is the most common flux for most ferrous and non-ferrous metals. It does not, however, dissolve oxides of aluminium and lead.
See: **annealing, dip, enamel, glaze, oxidation, pickle, solder, weld.**

corrosive symbol.

Focus

Pronounced: FO-KUS (*o as in go, u as in under*)

Origin

From the Latin *focus* meaning *the hearth* or *fireplace*; that is the main point of entry or comfort in a house, and the point to which people are drawn.

Meaning

The focus or focal point of a piece of work is that area to which the eyes are drawn most strongly. A designer may draw a spectator's attention to a particular part of a work (the focal point) by the use of contrasting shapes, vivid colours, changes in size, spotlighting, changing of values, by the addition of detail or even by distortion. To focus in photography is to adjust a camera lens so that a clear, sharp picture is seen through a **viewfinder** or on a ground glass **focusing screen**.

Focal length is the distance between a camera lens and the film when the camera is focused at the infinity mark. A **focus magnifier** is a device to allow a photographer to focus as accurately as possible. If a lens is focused for a specific distance, a subject can be behind or in front of that distance and still be in focus, so that a sharp image of the subject can be produced. This zone is called the **depth of field**. This should not be

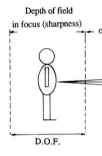

confused with the **depth of focus**, which is the amount a lens can be focused backwards or forwards and still produce a sharp image of a subject at a set distance away. **Zone focusing** is a simple method of setting the lens. Instead of setting a lens for a specific distance in metres, the camera has three broad categories of distance (*zones*), such as *landscape, group,* and *portrait*.

Associations

See: **distortion, foreshortening, laser, lens, perspective.**

Foreshortening

Pronounced: FOR-SHORT-EN-ING (*o's as in port, e as in token*)

Origin

Fore is an abbreviation of *before* or *afore* meaning *in front of. Shortening* originates from Old English meaning *to make smaller.* That is something immediately in front of you that appears to get smaller.

Meaning

Foreshortening describes an object that is at an angle to the plane which appears to become narrower and less clear as it recedes. It is a perspective effect in painting and drawing whereby objects or persons appear

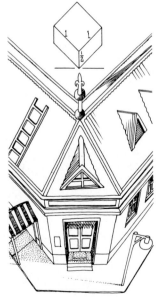

to be smaller than they really are. The spectator readjusts the objects or persons into their correct proportions. In foreshortening, the picture does not give a usual or characteristic view of the objects or persons. Geometrically, every projection involves foreshortening to some extent. A common example of foreshortening is where a figure in a picture stands with arms outstretched towards the viewer. The hands appear very large, the arms very short.

Associations

See: **angle, distortion, focus, perspective, plane, projection, proportion.**

Forge

Pronounced: FORJ (*o as in more*)

Origin

From Middle English *forgen* meaning *to build, to construct.*
Forging was one of the first metalwork skills, which was common over 6,000 years ago. In the 11th century metalworkers used the **trip-hammer** for forging and heavy hammers were raised by machines controlled by water-wheeels. In 1795, an Englishman, Joseph Bramah, invented the hydraulic press from which the forging press was developed. The invention of the **steam hammer** by a Scot, James Nasmyth, in 1839 revolutionised forging and led to the development of the **drop forge**.

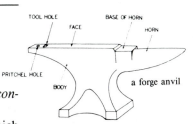

a forge anvil

a forged copper collar

Meaning

To forge is to shape metal by heating it to red heat in a fire and then hammering it, usually on an **anvil**. A **forge** is also a **smithy**, formerly called a blacksmith's **hearth**. In industry today, fairly small objects, like crankshafts, are forged by a **drop forge**, but massive objects, such as huge turbines and propeller shafts, are shaped by hydraulic forging presses which can exert pressures up to 50,000 tons.
Forged work is wrought-iron which is shaped by forging rather than by casting or riveted work.

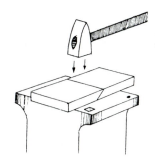

forge welding

Associations

See: **dress, hammer, hydraulic, mandrel, shape.**

Form

Pronounced: FORM (*o as in worn*)

Origin

From the Latin *forma* meaning *shape* or *configuration*.

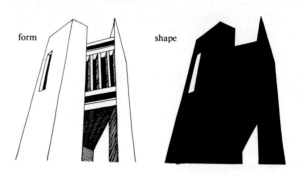

Meaning

Form is the shape, design and size of visual elements and their organisation and relationships in a work, in order to create unity in a composition. It is how the content of a work is revealed visually. Three-dimensional form is attained by (a) the **additive** method, involving *construction*, such as electroplating, laminating, sintering and powder metallurgy; (b) the **subtractive** method, involving *paring*, such as carving, milling or lathe work; (c) the **manipulative** method, such as, extruding, casting, moulding, forging, pressing and calendering.

A **former** is a specially-shaped piece of wood or metal (sometimes in two or more parts), which is used to form shapes by bending or pressing, as, for example, in laminating. A **form tool** is a cutting tool with a specific shape which can form a shape in wood in one operation.

An **artefact (also spelled artifact)** is something formed with skill and craft by a person rather than formed by nature.

In photography, form describes three-dimensionality, namely height, breadth and depth.

Associations

See: **composition, elements, format, hardening, laminate, malleable, pene, polygon, polyhedron, rectilinear, sculpture, shape, structure, symmetry, vacuum forming**.

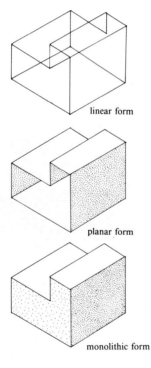

linear form

planar form

monolithic form

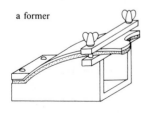

a former

Format

Pronounced: FOR-MAT (*o as in port, a as in cat*)

Origin

From the French *format* meaning *the form and size of something*, for example a book.

Meaning

Format refers to the size and general shape and composition of a design.
In photography, format refers to the height and breadth dimensions of a picture area.

Associations

See: **composition, drawing, form, hammer, mechanical drawing, technical drawing.**

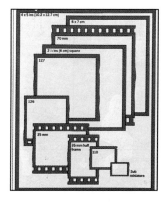

Fretted

Pronounced: FRETAD (*e as in let, a as in ago*)

Origin

From Old French *frete* meaning *trellis work*. A *trellis* is a structure of light wood struts (*laths*), so arranged that they cross each other. It is also called a *lattice*.
Elaborate carving and fretted work is characteristic of the furniture of Thomas Chippendale (1718 - 1779), who is recognised for his superb workmanship.

Meaning

Fretted describes the ornamental pattern made by the interlacing and crossing at right angles of bars of metal or laths of wood. A *chequered* effect is produced.
Fret work in building is a type of glazing in which diamond-shaped panes of glass, separated by lead strips, are formed into a design for a window. The term (also called **fret**) also applies to intricate, curved, open designs in wood where a **fret saw** is used, which is a handsaw with a long, narrow, straight, tapering blade. This saw is also called a **compass saw**, a **lock saw** and a **key saw**.

Associations

A **trellis** is a a structure of crossed-barred work made of either wood or metal which is used for supporting, among other things, plants in a garden.
See: **glaze, grid, jig, lath, saw.**

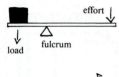
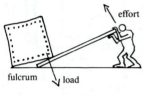
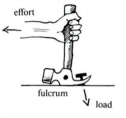

Fulcrum

Pronounced: FUL-KRUM (*u's as in fun*)

Origin

From the Latin *fulcire* meaning *to prop*, meaning *to support*, for example with a rigid pole.

Meaning

A **fulcrum** is the support on which a lever turns. It is also the exact point at which the lever is placed to get a **purchase** (meaning the most mechanical power, advantage or **leverage**), in order to raise or move something heavy.

The simple formula for the working of levers is: the *effort* multiplied by the *distance* from the *fulcrum* **equals** the *load* multiplied by the *distance* from the *fulcrum*. The greater the distance between the fulcrum and the point of making the effort, the easier the work is done.

Associations

See: **cybernetics, ergonomics.**

Furniture

Pronounced: FUR-NI-CHA (*u as in burn, i as in ink, a as in ago*)

Origin

From the French word *fourniture* from *fournir* meaning *to equip* or *to supply*. It came to mean those items which equipped a house or other building.

The development of furniture reflects the development of science and technology. As new materials, machines and techniques became available, so new kinds of furniture and new styles were developed. For example, wood dominated as the material for making furniture until the 1920s when designers such as Marcel Breuer and Mies der Rohe, past students of the Bauhaus, produced tubular steel framed furniture and during the 1930s bent laminated wood furniture was constructed. During the 1940s American designers Charles Eames and Eero Sarrinen produced chairs made from moulded plywood which were fixed by synthetic adhesives which had recently been discovered. In the same decade plastic furniture was first constructed.

Meaning

Furniture describes the moveable contents of a building or room, such as chairs, cabinets, beds, tables etc. or fixed items, such as built-in wardrobes or cupboards. Furniture is made by cabinet-makers, joiners, wood machinists and chairmakers.

It refers, too, to **accessories** (meaning *additional things*), which are minor fittings and attachments, such as door handles and locks.

In workshops, it refers to **tote boxes** (metal boxes used for storing small parts), tool racks and other storage devices.

Associations

To furnish is to fit up, supply, or equip a house or room with the necessary appliances, especially moveable furniture. Parts of a piece of furniture which are not made from wood are called **hardware** (e.g. nails, brads, screws, hinges, catches).

See: **carcass, fasteners, joinery, inlay, laminate, marquetry, mortise, particleboard, plinth, stile.**

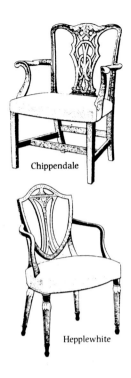
Chippendale

Hepplewhite

Fusibility

Pronounced: FUZA-BILITY (*u as in due, a as in ago, i's as in pin, y as in duty*)

Origin

From the Latin *fusus* meaning *poured* or *melted* from *fundere* meaning *to pour, to melt*.

Meaning

Fusibility is the capacity for a substance to be **fusible**; that is to be able to melt. The **fusability** of a metal is its readiness to change from a solid to a liquid form.

To fuse means to melt, liquify or dissolve under intense heat. **Fusion** is the melting and blending of substances. A **fuse** is a safety device in electrical circuits. When a circuit is overloaded with excess current **fuse wire** in a fuse melts and the circuit is broken.

Fuse welding is a process of bonding two pieces of similar metal by fusing an edge of each of the pieces to form one piece of metal.

Associations

See: **circuit, weld.**

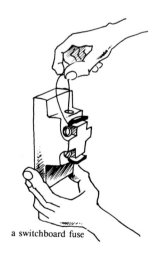
a switchboard fuse

Gauge

Pronounced: GAJ (*a as in late*)

Origin

Probably from an Old French word *jange* which became *gauger* meaning *the action or result of measuring*.

The micrometer gauge (*micro* being a prefix meaning *one millionth of*) was invented by an Englishman, William Gascoyne, in 1640. This was improved in 1848 by a Frenchman, Jean-Laurent Palmer, who produced the first micrometer which could measure to 0.05 millimetre. The **vernier gauge** was invented by a Frenchman, Pierre Vernier (1580 - 1637).

Meaning

A gauge is an instrument for the measurement of dimensions, volume or pressure.

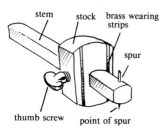

marking gauge — stem, stock, brass wearing strips, spur, thumb screw, point of spur

In woodwork, a **marking gauge** is used to mark out a parallel line to a face or to gauge the width or thickness of something; a **mortise gauge** is used to mark out two parallel lines to set out a mortise and tenon joint; a **cutting gauge** is used to mark out a line cut across a grain parallel to an end. It is also a tool used to measure or control the depth and angle of a hole drilled into wood.

In metalwork, gauge is a term for the grading of sheet-metal less than 3.125 mm. thick. The higher the grade number, the thinner the metal. Gauge 18 is for metal 1.3 mm.; gauge 22 is 0.7 mm. metal. It is used also to measure the spaces between metal parts (*feeler gauges* and *sheet and wire gauges*), to determine the type of thread and the pitch of various screw threads (*screw pitch* or *thread pitch gauge*), to measure the angle for screws of cutting tools (*screw cutting gauge*), to measure radii (*radius gauge*) and to measure the internal diameters of objects (*telescopic bore gauge* or *small hole gauge*). A *snap gauge* is a gauge preset to measure correct sizes in production work.

feeler gauges

A **drill gauge** is a piece of steel plate drilled with holes of varying sizes and marked so that the size of a drill can be readily known by testing it in an appropriate hole. A **taper gauge** is used for testing the accuracy of internal and external tapers.

The term gauge also describes the thickness or diameter of electrical wire. The **Standard Wire Gauge** (S.W.G.)

is a system which uses numbers to indicate the gauge of wires - the lower the number the thinner the wire. For example, 36 SWG wire is 0.2 mm and 18 SWG

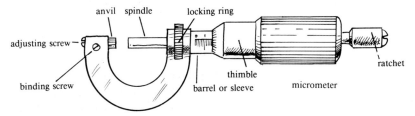

wire is 1.2 mm in diameter.

Associations

A **gauge wheel** is used for measuring the thickness of sheetmetal or the diameter of a wire. **Micrometers** are precision instruments which can measure directly to 0.01 mm. There are three kinds: **outside micrometers** to measure external dimensions, **inside micrometers**, which measure between internal surfaces and **depth micrometers**, which measure the depth of holes and grooves and the distances between the shoulders of objects.

See: **drill, face, grain, indexing, mill, mortise, screw, taper, template, threads, tolerance.**

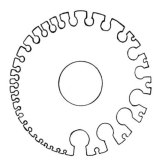

a gauge wheel

Geometric

Pronounced: JEE-O-MET-RIK (*ee as in see, o as in go, e as in let, i as in brick*)

Origin

From the word *geometry* meaning *the science of the size of things in space* (e.g. lines, surfaces and solids). It is from the Greek *geo* meaning *earth* and *metria* meaning *measurer*.

Meaning

Geometric means created by the precise mathematical laws of geometry. Usually simple shapes such as circles, rectangles and triangles are used, although more complex shapes are sometimes used. Geometric tools are used for measuring, and marking or setting out. They are: rulers, a straight edge, a winding stick (a pair of straight pieces of timber with parallel edges), wing compasses (dividers with pointed legs), callipers, a try square,

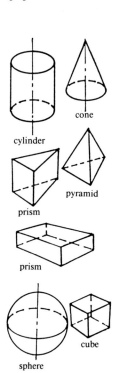

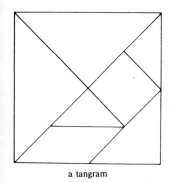
a tangram

a sliding bevel, a mitre set, and a variety of gauges.
A **tangram** is an ancient Chinese puzzle which can be used to form geometrical sculptures. A large square is divided into seven parts, comprising a square, five right-angles triangles and a parallelogram. From these seven parts an infinite number of shapes can be created.

Associations

See: **angle, computer graphics, drawing, elevation, form, Golden Section, line, mechanical drawing, perspective, plane, polygon, polyhedron, projection, rectilinear, scale, sculpture, shape, space, symmetry.**

Gilding

Pronounced: GILD-ING (*i as in ink*)

Origin

From the Old English *gyldan* meaning *gold*. When gold was first used is not known for certain, but gold utensils found in Iraq show that things made with great craft were being made in gold more than 5000 years ago. The process of gilding has been used for thousands of years in China, Japan, Egypt and Ancient Greece and Rome.

Meaning

Gilding is the technique of applying real gold in either very thin leaf or dust form to objects (e.g. to porcelain). Gold, unlike bronze or paint, does not lose its sheen, even when left outdoors and is subjected to changing atmospheric variations. It can be bought in books consisting of very thin gold leaves about nine inches square or in gold dust. Gold is the most malleable and ductile of all metals When it is beaten into thin sheets, it is called **foil** and in the very thinnest sheets it is called **leaf**. The leaf or foil is applied to objects using an adhesive called a **mordant**, which *bites* into the leaf to provide a surface which makes the gold adhere. Gold leaf is usually **tooled** to provide textual variations instead of the plain, flat texture of the gold leaf.

In metallurgy, **gilding metal** is a copper-zinc alloy containing 95% copper and 5% zinc.

Associations

The term **gild**, or **gilt** as it is sometimes called, also refers to silver as in **silver-gilt**. **A gilder's tip** is a brush which when brushed through a worker's hair produces enough static electricity for delicate leaves of gold to be picked up by the worker to be placed carefully on an object which is being decorated. Gold is measured in **carats**, a carat being a twenty fourth part. Twenty four carat gold is 100% gold; eighteen carat gold is 75% gold and nine carat gold is 37.5% gold. **Noble metals** refer to gold, silver and platinum, which do not change despite atmospheric changes.

See: **ductility, etching, malleable, mordant, noble metals.**

Glaze

Pronounced: GLAZ (*a as in late*)

Origin

Glaze is a Middle English form of *glass*.

Meaning

Glazing is when the particles of abrasive material on a grinding wheel or sanding disc lose their sharpness and begin to clog.

To glaze refers to the fitting of sheets of glass into a window or door frame by a **glazier**. Nowadays, sheets of polycarbonate are used as windows where there is danger of breakage, as the plastic is clear, transparent and almost unbreakable.

In photography, to glaze means to give a glossy sheen to photographic paper. Wet prints are rolled (using a rotary glazer) or squeezed (using a flat-bed glazer) against a polished chromium-plated sheet and then heated for about five minutes until the prints are dry. Nowadays, plastic coated (resin coated) papers produce a mirror gloss without glazing.

Glaze also refers to the placing of layers of transparent film of paint onto the ground or colour masses of oil paint in order to achieve a depth and brilliance of colour. Generally, a darker colour is placed over a lighter one. A glaze must be thoroughly dry before another layer of glaze is added. Tiles and bricks can be given a glazed, glassy finish by spraying them with a glazing material before they are fired in a kiln.

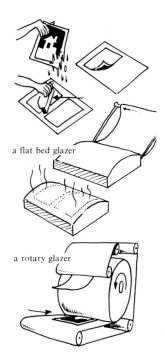

a flat bed glazer

a rotary glazer

Associations

See: **abrasive, annealing, calender, develop, flux, fretted, grind, honing, kiln, matt, paint, plastic, varnish.**

Golden Section

Pronounced: GOL-DAN (*o as in go, a as in ago*), SEK-SHUN (*e as in let, u as in bonus*)

Origin

The term *Golden Section* (also called *Golden Ratio, Golden Mean* and *Golden Number*) was first used in the nineteenth century but the concept goes back to Ancient Greek times and was used by the famous Greek mathematician Euclid in about the year 300 B.C.. It was applied first to architecture in the first century B.C. by Vitruvius in his treatise "*De Architectura*".

Meaning

Golden section

Golden section is the name given to a geometric proportion or ratio where a line is divided so that the smaller part is to the larger part as the larger part is to the whole. This works out at a ratio of about 5 to 8 (1: 1.618). It is expressed algebraically as $a:b = b:(a+b)$. It has been used extensively in drawing. It is said to express the secrets of **visual harmony** and is regularly found in nature, for example, the widths of curves upon a shell's spiral, the number of leaves on a plant's stem, the proportions of the human body, or the structure of a crystal.

Le Corbusier, a French designer and architect, devised what he called his **modular** system of architectural proportion based on the Golden Section.

The **Fabonacci series** (or *summation Series*) has a ratio where each number is the sum of the two numbers before it, e.g. 1, 2, 3, 5, 8, 13.

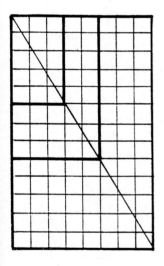
Fabonacci Series

Associations

See: **geometric, mechanical drawing, proportion, technical drawing.**

Gradation

Pronounced: GRA-DA-SHUN (*a's as in late, u as in bun*)

Origin

From the Latin *gradus* meaning *a step*. A grade is one "step" as you advance and one level of something (e.g. creative ability, or tone in colour) on a total graded scale. The Latin *ation* means *forming* or *showing*, so gradation is *how grades are shown*.

Meaning

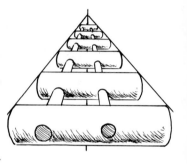

Gradation describes steps, stages or degrees of gradual, progressive change in an orderly fashion. Gradation can be shown by regular changes in colour, shape, size, space, direction, position, tone, texture, and gravity or a combination of some of these. Speed of change is decided by the number of steps (grades) in a gradation process. To **grade** is to arrange things in gradation. Gradation describes the lines on measuring tools and the dials of machines to indicate units and parts of units to be measured. **To graduate** is to divide a piece of metal into equal parts by incising lines (graduation marks) into the metal.

In photography, a **graduate** is a plastic or glass measuring cylinder with graduated marks etched on the side of the vessel. They are used for measuring precisely amounts of chemicals to be used in photographic processes.

a graduate

Associations

In photography, a **grey scale** is a gradation of grey tones, starting with white and going to black. It is used as a test or guide in photography.

See: **cross hatching, half-tone, indexing, mechanical drawing, negative, pattern, technical drawing, tone.**

Grain

Pronounced: GRAN (*a as in plane*)

Origin

From the Latin *granum* meaning *grain, seed, small kernel*. It came to mean *the smallest particle of anything* and then the arrangement of the particles which made

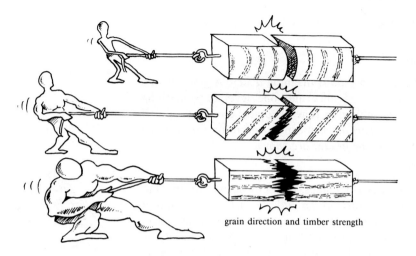
grain direction and timber strength

up a substance, e.g. the particles in wood, stone, leather or flesh.

Meaning

Grain describes the lines of fibre particles which are arranged in a clear direction and pattern in wood, stone and leather. Grain also describes the **granular**, mottled, texture of the surface of some substances, which are often rough.

The **end grain** in wood is the grain which is produced by cutting across the face of a plank of timber. **Cross grain** is where wood fibres deviate from the pattern of parallel fibres. **Winding** describes the diagonal twisting across the grain of timber.

In photography, the grain of a photograph is its texture, produced by the size and spacing of the particles of silver crystals or dyes in the emulsion on the photographic film or paper.

Graining is the uniform scratching of the surface of a metal for decorative purposes.

In painting, graining describes the brushing or combing of paint while it is still wet to produce an effect of grained wood.

Associations

A **granule** is a small grain. In plastics, it refers to small, cylindrical pieces of plastic material often used in machine-moulding processes, such as injection moulding.

Granulation means to roughen the surface of a substance by breaking it up into small grains or masses. Metals can be **granulated** by pouring them while hot through a sieve into water. In metalwork, grain is another term for *crystal*. The **granular structure** of metals can be seen under a microscope.

See: **bromide, cleavage, distortion, emulsion, figure, fissile, gauge, injection moulding, negative, knot, matt, metallurgy, paint, plastic, rip, texture.**

Grid

Pronounced: GRID (*i as in bit*)

Origin

From *gridiron* or *griddle* which are early English words for a cooking utensil, consisting of intersecting metal bars with spaces between them. It was used for boiling or grilling food over an open fire.

Grids (as a network of vertical and horizontal lines) were used by Renaissance (approximately 1400 to 1650) artists for scaling their sketches and cartoons (full-size, preparatory drawings) which would be used in murals. Grids were also used extensively by early typographers in the 15th century to design letters and to lay out printed pages.

Meaning

A grid is a system of parallel lines which cross each other at even intervals to create equal rectangles or squares (as on *graph paper*). Often a grid is on transparent material so that an image under the grid can be seen. It can be used for arranging printed matter or images in a regular pattern, to check whether lines are parallel, to identify positions on a picture or photograph, to measure distances and to enlarge or reduce the size of a drawing. A grid with lines spaced 15 mm apart will increase the size of the original by a half. A 50 mm grid will double the size of the original drawing. Grids are sometimes used for large, complex engineering and technical drawings and building plans where it is necessary to divide the total drawing into its constituent parts. The various parts of the drawing are called **zones**, each of which has a **grid reference**.

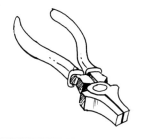

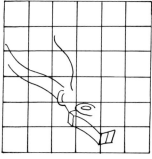

A grid is also an electrode which has one or more

openings for the passage of electrons or ions, and the base on which plates are set in storage batteries.

A grid can also be used as a kind of filter (a grating) to prevent unwanted materials from entering, for example, a pump inlet or a conduit.

Associations

Graticule is an alternative word for grid. **Profile paper** is drawing paper which is covered by a grid, with lines spaced according to what scale is used.

See: **drawing, dress, electrode, filter, fretted, mechanical drawing, profile, proportion, scale, sketch, squaring up**.

Grind

Pronounced: GRIND (*i as in find*)

Origin

From Old English *grindan* meaning *to gnash the teeth*. Grinding to pulverise substances and to polish materials has existed since the beginning of the earliest peoples on earth.

The earliest rotary grinder is shown in the Utrecht Psalter (Germany) which dates from about 816. Leonardo da Vinci (1452-1519) made sketches of a disc grinder.

As industrialisation developed in the 19th. century, it became essential to produce, cheaply, accurately-machined surfaces for a range of machines. The invention of precision grinding machines made this possible. Most of the developmental and production work was done in the U.S.A (e.g. by James Wheaton, James Whitelaw and Samuel Darling during the 1830's). By 1896, carborundum-wheels were marketed and the precision grinder became a common production tool.

Meaning

To grind a substance is to reduce it to small particles or powder by crushing it. It also means to wear down a substance or to sharpen it by creating friction by vigorous rubbing with a material much harder than the substance.

Grinding on a **grindstone** is used to reshape cutting materials which have been chipped or whose cutting edge has been rounded. It is the first stage of a sharpening process. The second or fine-sharpening stage fol-

grinding a round-nosed tool

lows with **honing** on an **oilstone**.

A **grinding machine** is a power-driven device which can machine interior and exterior surfaces, make and finish holes, polish screw threads and can produce finishes to high tolerances. The **universal grinding machine** has a swivel table, swivel headstock and swivel head. It can be used for internal and external cylindrical grinding and face and surface grinding.

The main types of abrasives used with metals are aluminium oxide (on mild, carbon, alloy and high-speed steels, wrought iron and hard bronze) and silicon carbide (on cast iron, brass, aluminium, copper and materials of low-tensile strength).

The **grinding allowance** is the amount of material which is left on a workpiece to allow for finishing by grinding.

Associations

A **grindstone** is a thick disc made of quarried natural sandstone which rotates to grind a substance, in order to reshape, sharpen or polish it. The stone is kept wet from a reservoir of water to reduce friction heat and to wash away particles of stone and metal. **To ground** also means to grind. **Mulled** means ground to a fine powder. **Blueing** of a cutting tool which is being sharpened on a grinder indicates that the blade is overheating, which can soften the metal.

Ashing is the grinding of plastic using a *buffing* process.

To comminute is to grind or pulverise solids.

See: **abrasive, buff, facet, glaze, honing, lap, mill, plastic, plinth, shape.**

Groove

Pronounced: GROOV (*oo as in soon*)

Origin

From The Dutch *groeve* meaning a *trench, channel or furrow*.

Meaning

A **groove** is a furrow or recess (e.g. square, rectangle or v-shaped) made with a cutting tool (e.g. a chisel, router plane, plough plane, or a milling machine) into a substance, e.g. wood, metal or plastic. A groove usually runs lengthwise rather than across the material being worked on. Often it is made to direct movement or to make a joint with another piece of material which is ridged. It is said to have a **tongue** which fits into a furrow to produce a **tongued and grooved joint**. Splines are grooves cut into the end of a spindle or shaft.

Associations

To plough is to cut a groove in wood. A **glyph** is a short, vertical groove.

See: **chisel, cross section, flute, intaglio, joint, key, knurling, mill, mortise, plane, rebate, rout, scoring, serration, spline, striation.**

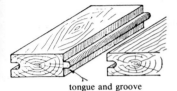
tongue and groove

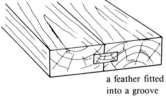
a feather fitted into a groove

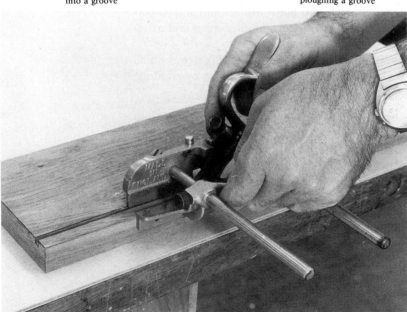

ploughing a groove

Half Tone

Pronounced: HAF-TON (*a* as in *far, o* as in *bone*)

Origin

The first photographs were made by a Frenchman, J.N.Niepce, in 1822 and his invention was developed by an Englishman, William Henry Fox Talbot, who published the first illustrated book with photographs, called *The Pencil of Nature* in 1844. Talbot discovered the principle of half-tones. The first person to reproduce photographs in printers ink was a Swedish engraver Carl Gustaf Wilhelm Carleman in 1871, who used line half-tones (rather than dot half-tones which came later). The photo-mechanical process was invented by Frederick E. Ives of the U.S.A. in 1880. The first half-tone photograph in a newspaper appeared in the *Daily Graphic* in New York. The process allowed photographs to be reprinted at the same time as text, instead of as a separate operation, which meant that photographs could be used extensively in all printing operations, especially in newspapers.

Meaning

A printing press (e.g. letterpress or litho-press) can lay down ink on paper only of one density. It cannot make the ink in one area of a page darker than another area. Therefore, in order to produce the range of tones present in a photograph for a printing process, a new technique had to be devised. As the new process did not produce continuous tones (i.e. the gradual merging from black to white) but had areas of tone and areas without tone, it was called **half-tone**. In half-tones, an image on a photograph is broken down into lines of separate dots, or elliptical dots, where the larger the dots the darker is the image. The areas of least density are the **highlights**; those with the greatest density produce shadows. To produce the dots on a photograph, a **half-tone screen** is used, which consisted in the past (and to some extent today) of a sheet of glass onto which fine, intersecting lines are etched, but which nowadays is more often a contact screen made on a film base. The screens vary in the number of lines they have on them, ranging usually from 20 to 80 lines to the *centimetre* (but they can exceed this). The greater the number of lines, the finer is the screening. Under nor-

50% square dot screen

50% elliptical dot screen

half-tone screens

Concentric circle

Wavy line

Parallel line

Brick

marks of quality

LONDON

EDINBURGH

BIRMINGHAM

SHEFFIELD

mal viewing conditions the eye is not able to see the dots which make up the images which vary from light grey to almost black. Very fine screens (in excess of 150 lines) can be used only when top-quality art paper is being used for reproductions. The half-tone screen is placed behind the lens of a copy camera which makes negatives for printing plates.

A large number of half-tone "contact" screens of various designs and patterns is available commercially, which may be used to produce specific textural effects on an image. In engraving, half-tone effects can be produced by fine *cross-hatching* or by the use of dots as in photography.

Associations

See: **bromide, cross hatching, gradation, line, negative, positive, tone** (see *colour*), **engraving**.

Hallmark

Pronounced: HAL-MARK (*a as in fall, a as in car*)

Origin

From Old English *healle* meaning *hall*. The word *hall* took on many meanings, among which were places for conferring degrees, professional qualifications and licences.

Meaning

A **hallmark** is a mark or symbol placed on articles of gold, silver and platinum at various assay offices in the United Kingdom to indicate their true value and the fineness of the metal. Goldsmith Hall in London is the most famous of the assay offices. **To assay** is to test metals or ores for purity and quality.

Associations

See: **symbol**.

Hammer

Pronounced: HAMA (*1st a as in man, 2nd a as in ago*)

Origin

From the Old English *hamor* meaning *the back part of*

an axe or *an instrument for hitting and breaking hard things*. A hammer is a **percussion tool**. *Percussion* is from the Latin *percussio* meaning *to shake or strike thoroughly*.

Hammers with handles which were attached by thongs of leather (i.e. the *haft*, which increased the force of a blow) were used during the Middle Stone Age about 8000 B.C.. With the discovery of copper smelting in about 4000 B.C., the Ancient Egyptians used hammers with copper heads and about 3500 B.C they were cast in bronze. The claw hammer, with holes in the hammer head to take a handle, came into use during Ancient Roman times.

The first steam hammer was invented by James Nasmyth of Scotland in 1839 and was developed in France in 1840.

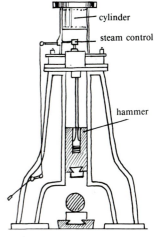

James Nasmyth's steam hammer

Meaning

Hammers, which are available in a variety of forms, are named according to the shape and size of their steel head or to their function. For instance, the **claw hammer** which is used where a fairly heavy tool is needed to drive in large nails, has a claw on one side of its head to extract nails. The **engineer's ball pene hammer** is a general-purpose hammer and is also used for riveting. The **warrington cross-pene hammer** is a general-purpose hammer for relatively light hammering and its cross-pene is used for starting small nails and brads. Hammers with variously-shaped penes are used for shaping sheetmetal work and for hammer marking, such as the **bullet head hammer**, the **repoussé hammer**, the **planishing hammer** and the **blocking or forming hammer**.

Soft-faced hammers (with heads of such materials as raw hide, plastic, copper, lead or brass) are used for hammering the surfaces of soft metals to prevent the surfaces from being marked or damaged. They are also used to strike a necessary heavy blow where a steel hammer would damage the workpiece.

Other percussion tools are **mallets**, which have large wood, rubber or hide heads. They are used in carpenters' assembly work, for striking chisels (e.g. a **carpenter's or joiner's mallet**) or for forming (e.g. **flat faced mallet**) and shaping sheetmetal (e.g. a **bossing or hollowing mallet**).

warrington cross pene hammer

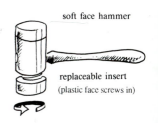

soft face hammer

replaceable insert
(plastic face screws in)

Associations

A **veneer hammer** has a rounded brass strip in its head, which is pressed down and moved in a zigzag movement to flatten veneers and force out air bubbles and surplus glue between veneers.

See: **burr, drawing, dress, ductility, emboss, forge, form, hardening, impression, jack, malleable, pene, pitch, plane, planish (in plane), repoussé, rivet, shape, swage, veneer.**

Hardboard

Pronounced: HARD-BAWD (*a's as in car, aw as in law*)

Origin

Hardboard is a 20th. century word coined to describe stiff boards of compressed and processed wood fibre.

Meaning

Hardboard is a relatively cheap and versatile material made from compressed hardwood fibre pulp which is bonded under heat with an adhesive. It is also called **fibreboard**. It has no grain and can be planed or sawn in any direction without splintering. It can be bought in boards in a wide range of sizes about 3mm to 12 mm thick and the boards can be patterned to imitate leather, wood grain or various textures. It is used for cheap carcass backs, drawer bottoms, inexpensive wall panelling and partitioning, pelmets and a variety of display boards. Special hardboard nails are used with the material.

Associations

See: **adhesive, carcass, joint, laminate, nails, pattern, plane, veneer.**

Hardening

Pronounced: HARDAN-ING (*1st a as in car, 2nd a as in ago*)

Origin

To harden and *hardening* derives from *hard*, which is from Old English *heard* meaning *strong* and *firm, unyielding, solid, not easily cut.*

Meaning

Hardening is a process of carefully-controlled heating and cooling of iron-based alloys to produce a hardness and tensile strength in the metal greater than that before the process. The cooling may be in water, air or oil, depending upon the composition and size of the metal. If tool steel is heated to specific temperatures, chemical reactions occur in the iron and carbon in the steel where new carbides of iron are formed. Different heat treatments can determine a metal's hardness, toughness, brittleness or elasticity. When steel has been hardened it tends to be brittle and this brittleness is reduced by **tempering**, which toughens the metal.

In photography, hardening describes the process where a chemical agent (e.g. chrome or potassium alum) is used to harden gelatine emulsion on a film to make it insoluble and permanent and more resistant to scratching.

Case hardening is a process of heating low carbon steel in contact with a carbon substance so that the steel absorbs carbon and has a high-carbon steel *"skin"* to a depth of several thousandths of a centimetre. Ferrous alloys heated in molten cyanide become case hardened. The core of the metal remains relatively soft but the outer *case* is hard. Bicycle spindles, ball bearings and roller bearings, some screws and gear wheels are often case-hardened.

Case hardening can also occur during the kiln-drying of timber. The exterior, or *case* of the timber dries out before the interior or core begins to shrink. The result is that the wood warps or is difficult to saw and work.

Flame hardening is a process of heating steel with the flame from an oxy-acetylene torch and then quenching the steel.

Metals harden as they are being hammered and worked. This process is called **cold working** or **cold forming** and results in what is known as **work hardening**.

Associations

There are various methods of non-destructive testing of the degree of hardness of metals. *The Brinell test* is operated by pressing a 10 mm hardened steel ball under a load of 3000kg into the surface of a metal and then measuring the diameter of the impression made with a microscope. *The Rockwell test* is operated by inserting

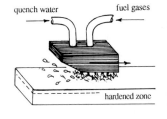

Flame hardening of flat plates

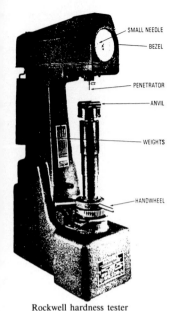

Rockwell hardness tester

a *penetrator* (e.g a conical diamond) into the metal at a known pressure and then measuring the depth of penetration. *The scleroscope Test* is operated by dropping a small diamond-tipped hammer from a fixed height onto a metal and then measuring the amount of rebound of the hammer from the metal surface. All these tests use *hardness tables* to obtain the hardness value of a metal.

The hardness of a metal is determined by (1) its resistance to indentation, (2) the stiffness or temper of the wrought material, (3) the ease or difficulty in machining the metal.

The hardness of minerals can be tested using a **Mohs Scale**, which has numerical gradations. Soft talc has a hardness of 1; diamonds have a hardness of 10.

See: **alloy, composition, ductility, emulsion, form, gradation, hammer, resin, shrinkage, temper, tensile, warp.**

Hasp

Pronounced: HASP (*a as in cat*)

Origin

From Old English *haesp* meaning *a clasp* or *hook*.

Meaning

A **hasp** is a clasp, which is a type of fastener. Usually the hasp is hinged and passes over a staple (a U-shaped piece of metal) and is secured by a metal pin or a padlock. The unit is commonly referred to as *hasp and staple*.

Associations

See: **fastener, fixing, hinge**

securing a hasp & staple

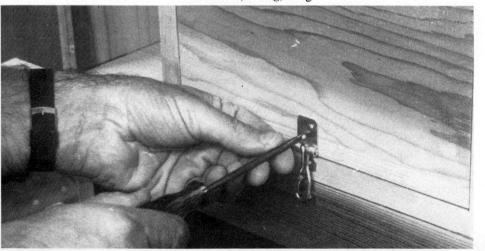

Haunch

Pronounced: HAWNSH (*aw as in saw, sh as in ship*)

Origin

From the Middle English *hanche* meaning *haunch, the fleshy part of the buttock and thigh*. This part of an animal was curved and rounded.

Meaning

The **haunch** is that part of an arch between the crown (the centre at the top) and the pier (the point where the arch begins to curve). It is also a recessed part of a **tenon** to prevent a mortise and tenon joint from twisting, particularly when pieces meet at a corner.

Associations

See: **joint, mortise**.

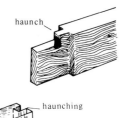
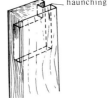

Haunched mortise and tenon

Honing

Pronounced: HON-ING (*o as in alone*)

Origin

From Old English *han* meaning a *stone*. The stone was also called a *whetstone*, as *whetten* meant *to sharpen*.

honing

125

Meaning

Honing describes a *fine* sharpening process, using a **hone**, which is a thick stone of natural or artificial (e.g. silicon carbide) fine grit, which has powerful abrasive effect on many metals. Honing usually occurs after the grinding of a tool. During the process, a cutting fluid (See: **coolant**) must be used on the stone and the tool being honed, to lessen the heat of friction and to wash away particles of stone and metal which could *glaze* (clog) the stone and make it ineffective as a sharpener. The cutting tool is held at an angle of 25° to 30° to the hone and is rubbed backward and forward until the tool has a bevel with a very sharp edge. As well as for sharpening tools, honing can be used to ensure a very close fit of parts where precision is necessary. The cylinders of engines are honed with very fine abrasives to obtain a high lustre.

Associations

Oil stones are used for honing tools to a fine edge.
See: **abrasive, bevel, coolant, fit, glaze, grind, lap**.

Hydraulic

Pronounced: HI-DRAW-LIK (*1st i as in fine, aw as in saw, 2nd i as in ink*)

Origin

From the Greek *hudranlikos* meaning *a water organ, a machine using piped water*. It comes from *hudor* meaning *water* and *aulos* meaning *a pipe*.

An Englishman, Joseph Bramah, invented the hydraulic press in 1795. Henry Maudslay, an English engineer, patented the hydraulic press for shaping metal in 1795. The lifting tool, the hydraulic jack, was invented for railway use by W.Curtis in 1838.

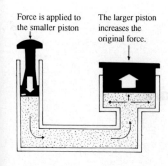

Force is applied to the smaller piston

The larger piston increases the original force.

Meaning

Hydraulics is the science which deals with liquids in motion (**hydrodynamics**) and the power and pressure they exert. It is the science of conveying liquids through pipes and other conduits as a source of power. Originally, hydraulics related to machines operated by water only but nowadays it refers to other liquids. For instance, a **hydraulic jack** is raised by forcing oil up against a piston or plunger; **hydraulic brakes** on motor vehicles act in a similar fashion, using brake fluid. A hydraulic press is used in the laminating of plastics.

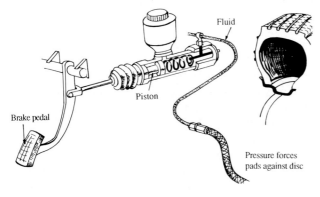

Associations

A **hydraulic press** is operated basically by the force created when water is moved from small to large cylinders. **Hydraulic mortar** is mortar which will harden under water. **Hydraulic glue** is glue which is partially able to resist the action of water or other moistures.
See: **adhesive, forge, jack, laminate, plywood.**

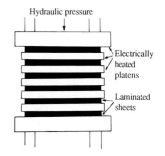

Hygroscopic

Pronounced: HIGRO-SKOP-IK (*1st i as in fine, 1st o as in alone, 2nd o as in stop, final i as in ink*)

Origin

From the Greek *hygros* meaning *wet, moist, humid* and *skopein* meaning *to look at* or *to examine*.

Meaning

Hygroscopic means tending to absorb and retain moisture from the air. A hygroscopic substance (called a **humecant** e.g. glycerin) can be added to paint to prevent it from drying too quickly.
A **hygroscope** is an instrument which shows variations in the humidity of the atmosphere.

Associations

A **hygrometer** is an instrument for measuring the moisture in the air. **Anhydrous** means free from moisture. **Hydrolysis** is the reaction of a material with water.
See: **accelerator, paint, shrinkage, seasoning.**

a thermo hygrometer

Imperial

Pronounced: IM-PEARI-AL (*i's as in ink, ea as in fear, a as in ago*)

Origin

From the Latin *imperialis* meaning *pertaining to the empire and the emperor. Imperium* meant *supreme power, command and order.*

In January 1826 the United Kingdom produced a compulsory uniformity of weights and measures within the British Empire (the "*imperium*"). This system of weights and measures was called the **Imperial System**. It contrasted with the **Metric System**, which was adopted by France on August 1, 1793 but which was not used as a system in that country until 1840. It gained universal acceptance for all scientific work. The United Kingdom and countries within its empire at the time preferred the Imperial system but they accepted the metric system in the 1960s. Now the metric system is used throughout the world, except in the Unites States and a few very small countries.

INCHES	MM	CM
⅛	3	0.3
¼	6	0.6
⅜	10	1.0
½	13	1.3
⅝	16	1.6
¾	19	1.9
⅞	22	2.2
1	25	2.5
1¼	32	3.2
1½	38	3.8
1¾	44	4.4
2	51	5.1
2½	64	6.4
3	76	7.6
3½	89	8.9
4	102	10.2
4½	114	11.4
5	127	12.7
6	152	15.2
7	178	17.8
8	203	20.3

Meaning

The Imperial System of weights and measures, which is still used to some extent today, has such basic units as *yard* (length), *gallon*, (capacity or volume) and *pound* (weight). The metric system has such basic units as *metre* (length), *litre* (volume), and *kilogram* (weight). The metric system is much easier to learn, understand and operate, as it is a decimal system, where calculations are made by multiplying or dividing by ten.

Impression

Pronounced: IM-PRESH-AN (*i as in dim, esh as in fresh, a as in ago*)

Origin

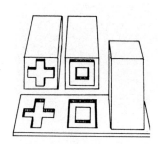

From the Latin *imprimere* meaning *to press upon, or force upon*, and later *impressio* meaning *emphasis* and *the act of stamping something.*

Meaning

An impression is the mark or dent usually made by punches, hammers or stamps.

Associations

See: **chased, die, emboss, engraving, hammer, repoussé, stamp, swage, symbol.**

Indexing

Pronounced: IN-DEX-ING (*i as in pin, ex as in text*)

Origin

From the Latin *index* meaning *a pointer, indicator, forefinger, sign, mark*. The Latin *indicare* means *to point out*.

Meaning

Indexing in engineering is to move a pointer on a measuring device through a fractional part of a complete turn or to some predetermined required position. It also means to subdivide a piece of work into equal parts or angles, using a **divider** (*a measuring compass*). An **indexing head** is a machine-tool attachment which allows workpieces to be rotated at required angles so that faces can be worked on.
An index is a pointer on an instrument showing a position on a scale.

Associations

The plural of index can be indexes or indices (pronounced *indi-seez*).
See: **gauge, gradation, lathe, scale.**

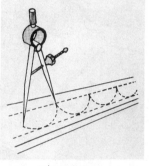

using a divider

Industrial Designer

Pronounced: IN-DUST-RI-AL (*i's as in pin, u as in dust, a as in cat*), DE-ZIN-A (*e as in delay, i as in sign, a as in ago*

Origin

Industrial designers have been employed in industry and commerce since the middle of the 19th century when the significance of producing machine-made articles which are both pleasing to look at and practical, efficient and effective was recognised. Schools of Design were founded in England in 1836. The development of the profession of Industrial Design is linked closely with the history of Industrialism. The term *Industrial*

Design was first used by the Americans Raymond Loewy and Henry Dreyfuss in the early 1900's to describe their professional activities in consultancies which provided services to American manufacturers. Peter Beehrens (1868 - 1940), one of the founders of the German Werkbund formed in 1907 (see: **Bauhaus**), is generally considered to be the first professional industrial designer.

Meaning

An Industrial Designer is a member of a product development team which is professionally responsible for enhancing the daily lives of people by designing products and environments which meet human needs in functional and aesthetic ways. Usually the industrial designer is involved in designing products which are made using machines and mass-production methods.

Associations

See the works of: Peter Behrens (Germany 1868-1940), Walter Teague (American 1883-1960), Wilhelm Wagenfeld (German 1900 -), Norman Bel Geddes (American 1893-1958), Henry Dreyfuss (American 1904-1972), Charles Eames (U.S.A. 1907-1978), Marcello Nizzoli (Italy 1887-1969), Walter Gropius (Germany 1883-1969).

See: **Bauhaus, computer graphics, drawing, design, ergonomics, mechanical drawing.**

Injection Moulding

Pronounced; IN-JEKSHUN- MOLD-ING (*i as in pin, e as in let, u as in fun, o as in solder*)

Origin

Injection is from the Latin *injectio* meaning *a throwing or flinging in*. Most words with *ject* im them (e.g. reject, project, subject) have a meaning where either throwing or pushing or thrusting out is involved. (See **Moulding**). The first commercial injection moulding machine was developed and patented in Germany in 1926 by Dr. Eichengrün. Automated production began in 1937, which resulted in many plastic commodities being plentiful and cheap.

Meaning

Injection means the act of throwing, forcing, driving or filling some cavity, especially with a liquid or solution. **Injection moulding** is one of the two **blow moulding** processes used in plastic production. The other is **extrusion blowing**.

In **injection moulding**, plastic powder or granules (of thermoplastic or thermosetting materials) are heated in a cylindrical container until they are soft and the plastic is then injected (nowadays usually by a screw mechanism) under pressure into a water-chilled mould. The injected plastic is cooled and then the shape produced is mechanically ejected.

injection moulding

photograph by kind permission of General Electric—Plastics

Associations

See: **blow moulding, die, extrusion, grain, moulding, thermoplastic, thermosetting, screw.**

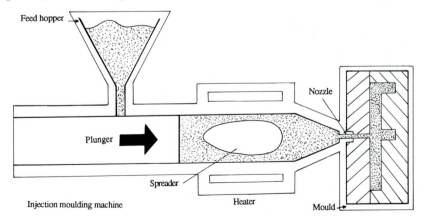

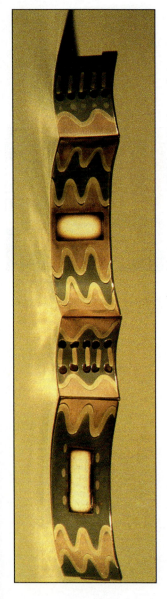
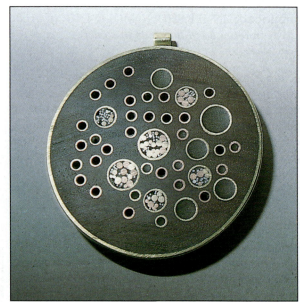

Inlay

Pronounced: IN-LA (*i as in pin, a as in plane*)

Origin

Lay is from Old English *lecgan* meaning *to put or place* and the preposition *in*. The modern derivation, *inlay*, came to mean *to embed, to ornament with something embedded in the surface*. Hepplewhite furniture (after George Hepplewhite, a famous English furniture designer who died in 1786) is characterised by inlaid mahogany work.

Meaning

Inlay is a decorative process where thin parts of the surface of wood or metal are cut out and another contrasting wood, metal or other material (e.g. ivory) is fixed into the cut-away area.

Certosino (pronounced Cherto-zeno) is inlay work where ivory or bone makes an attractive contrast on dark wood, such as ebony.

Associations

Damascene, **bidri** and **kuftgari** are types of metal inlay work. **Marquetry** is inlay work in wood.

See: **design, furniture, marquetry, pattern, rout**.

Intaglio

Pronounced: IN-TAL-YO (*i as in tin, a as in tar, yo as in yodel*)

Origin

From the Italian *intagliare* meaning *to engrave, etch* or *cut out*. In Italy an engraver is an *intagliatore*.

Meaning

Intaglio is the technique whereby a composition is hollowed out or scratched out from the material worked on. In **relief** work, the image required is above the material worked on; in intaglio work, the image is below the surface of the object. Intaglio includes the technique of **etching**. An intaglio is also an incised carving in a gem. It is the opposite of a **cameo**. Probably the commonest example of intaglio work is in a ring which is used as a seal.

Associations

Another term for intaglio is **diaglyph** and its opposite, the relief form, is called an **anaglyph**.
See: **chasing, composition, etching, groove, relief, tooling**.

Jack

Pronounced: JAK (*a as in black*)

Origin

The origin of *jack* is not known. It probably came from the Malayalan (South India) *chakka* meaning *something round* which became *a wheel* or *circle*.
A sketch exists of a screw-jack drawn by a French architect, Villard de Honnecourt, about 1250. In 1755, George Staghold and ten years later George Pickering invented a screw-jack with a gear-driven nut. During the 1800s many inventions were made to improve lever-and-screw jacks. In 1838, W. Curtis invented the first hydraulic jack.

Meaning

The word *jack* has many meanings. Those related to tools or equipment are: a machine for turning a spit to roast meat, a support on which to saw wood and a

Screw Jack

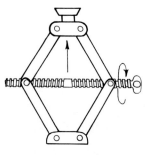

a scissor screwjack

screw (a jack-screw) for lifting heavy weights, such as support for buildings while they are being repaired and for lifting a car so that its wheels can be removed while they are being changed. A portable jack can be operated, so that a support mechanism is raised from a firm base, by winding or pumping using a lever action, by ratchet action, by a rack and pinion mechanism, by a screw mechanism or by hydraulic pressure. **To jack** is to raise up something.

Associations

A **jack hammer** is a powerful, hand-held pneumatic hammer-drill, used for breaking up rock and hard substances, such as road surfaces. A **jack plane** is a large, strong plane used for coarse joinery work, usually before finishing with a smoothing plane.
See: **hammer, hydraulic, plane, ratchet, screw**.

Jig

Pronounced: JIG (*i as in big*)

Origin

The word *jig* probably comes from the Old Norse word *giga* meaning a *fiddle* or the Old French *giguer* meaning *to hop or dance*. Both words imply moving in a lively but ordered manner.

Before the 1st century Romans used jiggers to make multiple clay objects, as did the Chinese. It was then applied to a horizontal lathe used to make flat ware, such as plates and dishes.

Meaning

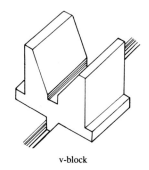

v-block

Angle iron to hold material for butt welds

A jig is an appliance which holds a piece of wood, metal or plastic work and guides the tools working on it, for example a dowelling jig for guiding a bit which is boring holes when making a dowel joint, or a welding jig for holding metal at a 90° angle during a welding process. A **bending jig** is a tool used for bending curved sections of metal by hand. A **jig saw** is a **fret saw** or **scroll saw** which is operated by a machine which moves up and down. It is used for cutting thin wood or plastic into ornamental patterns (e.g. *volutes*), particularly those which have interior and exterior curves. Its unusually narrow blades allows the cutting of very acute curves. A **jigger** is a machine which operates with a rapid up-

and-down, jolting movement. It is often used for separating or sifting materials.

Associations

See: **boring, dowel, fretted, lathe, pattern, saw, screw, weld.**

using a jig saw

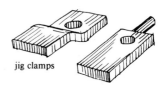

jig clamps

Joinery

Pronounced: JOYN-ARY (*oy as in boy, a as in ago, y as in duty*)

Origin

From the Old French *joigneor* meaning *to join, unite.*

Meaning

Joinery is the skill in joining wooden things together, It is carried out by a **joiner** who makes furniture, house fittings and and other woodwork which is lighter and more intricate than that made by a **carpenter**. A **cabinet maker** is a very skilled joiner, who usually specialises in the making of fine furniture.

Associations

See: **furniture, joint.**

Joint

Pronounced: JOYNT (*oy as in boy*)

Origin

From the Old French *joigneor* meaning *to join, to unite*

Meaning

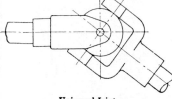

Universal Joint

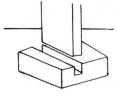

through housing joint

a dovetail joint

A joint is a point at which, or means by which, two parts of something being constructed are united, either rigidly so that no movement can occur or so that controlled and limited movement can occur, as with a hinge. The wedging of one piece of wood into a cavity of another piece is called **socketing**.

There are many wood joints which are used for specific purposes but they can be put into three broad categories: **widening joints** (e.g. butt and dowell butt joints), **carcass joints** (e.g. rebated and rebate with fillet joint) and **framing joints** (e.g. haunched mortise and tenon joint).

The **shoulder** of a wood joint is its base, which gives extra strength to the joint.

To Joint is to prepare a board (e.g. hardboard) which is to be joined to another by planing its edges.

Joints used in sheet metal fabrication are usually made by folded *seams*, which are then finished by soldering, riveting or a combination of both.

Associations

Halving is a method of cutting and joining pieces of wood where half the thickness of the face of one piece of wood and the remaining half from the back of the second piece are joined, so that their outer surfaces are flush. A **T joint**, a plate of metal shaped like a letter T, is used to strengthen a joint.
A **universal joint** is a *coupling* which allows the rotation of two shafts whose axes are not in a straight line.
See: **bead, burr, carcass, chasing, dowel, fabric, filler, fillet, fit, groove, hardboard, haunch, hinge, joinery, lap, mitre, mortise, notching, pinning, rebate, rivet, scarfing, seam, solder, spline, tenon, weld.**

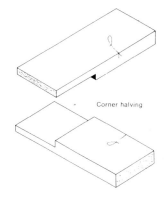

Corner halving

Kerf

Pronounced: KURF (*u as in surf*)

Origin

From Old English *cyrf* meaning *a cutting or carving* from *ceorfan* meaning *to carve or cut*.

Meaning

A kerf is a slit or groove made by cutting a material, especially the cutting of wood with a saw. It describes also the cut end of a felled tree.
Kerfing is the bending of solid wood or plywood by sawing kerfs part-way through the material (about two thirds of the thickness of the piece) to lessen the normal stresses made on solid materials when they are bent. Curved faces or edges can be made by this method. The strength of the material is reduced by kerfing, but where strength is not essential (e.g curved small doors or curved drawers) the technique is effective.

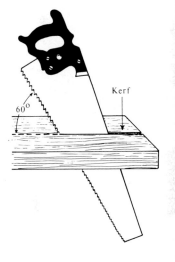

Associations

See: **cross section, groove, plywood, saw.**

Key

Pronounced: KEE (*ee as in see*)

Origin

From the Old English *caeg* meaning *an iron instrument for moving bolts of a lock forward or backward to fasten*

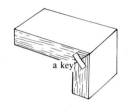
a key

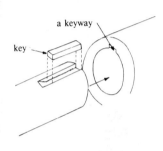
a keyway
key

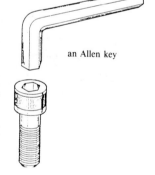
an Allen key

or unfasten. Key came to mean what opens up anything to reveal its importance.

Meaning

A key is piece of wood or metal inserted between two pieces of wood or metal to secure them, often at a corner joint (e.g. the strengthening of mitre joints by gluing slivers of veneer into saw kerfs at the joint). It is also a piece of wood let into another piece crosswise to prevent warping. Also, it is an insert in the seam of a mould which is made in two or more pieces, which when removed enables the mould to be taken to pieces without damage. In metal work, it refers to a small, removeable, (usually rectangular but also semi-circular) piece of metal which fits into slots in both a shaft and a hub to lock a gear or a pulley on a shaft, so that they will rotate together. The **key way** is a groove to receive a key. **Splines** allow a gear to move longitudinally along a shaft; keys do not.

It is also a mechanical device for making or breaking an electrical circuit (e.g. an ignition key), and an instrument for grasping screws, nuts etc. (for example to wind a clock).

To key a surface is to roughen it to assist the adhesion of materials to it. **To key** can also mean **to fasten, wedge or bolt** something.

Associations

An **Allen key** is one of a set of bars used for tightening grub screws or particleboard screws. It normally has a hexagonal (i.e. six-sided) cross-section. It is also called a **hex key**. A **woodruff key** is a half-moon shaped piece of metal which is designed to fit into a semi-circular groove in a shaft.

See: **adhesive, circuit, colour (values), groove, mould, mitre, particleboard, priming, seam, spline, veneer, warp.**

Kiln

Pronounced: KILN (*i as in pin*)

Origin

From the Latin *Culina* meaning *where the baking takes place*. The word *culinary* means *related to cookery*. The word kiln came to mean the container in which clayware is baked or *cooked*.

138

Kiln sites have been found during excavations in China which date back to 6000 years ago. High-temperature kilns to produce proto-porcelain existed in the Shang Dynasty in China from the 16th to 11th century B.C..

Meaning

A kiln is a high-temperature oven or furnace heated by electricity, gas, oil or combustible material such as wood. It is used to fire ("*bake*") pottery, to fuse enamel objects and for glass staining. It must be heavily insulated and various methods are used, including double refractory (*heat resistant*) walls, and refractory foam and filament fibres. There are many kinds of kilns, which are sometimes described by their shapes, e.g. *beehive kiln, tunnel kiln, hovel kiln* (like a bottle). A small workshop kiln is usually fired by electricity or gas. It is often a simple box construction weighing, depending upon the materials used, about 50kg. (opening at the top or in front), which has inside a length of resistance wire and refractory brickwork for insulation. It is easy to operate as the normal fuel supply (electric power) is easy to control and there are no waste disposal problems. It is readily installed and, if fitted with silicon carbide elements, will stand the strain of **reduction**. (See: **oxidation**.)

a gas kiln

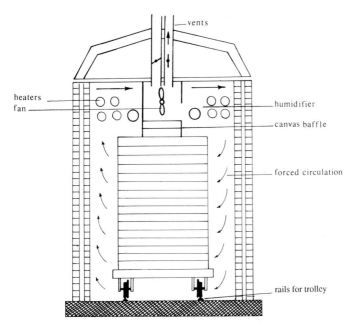

a seasoning kiln

A kiln is also a large building where timber can be stacked in such a way that hot air and steam can circulate around all the surfaces of the timber during a **seasoning** process. The temperature in the kiln must be monitored carefully to ensure the timber does not dry out too quickly, which can produce distortion in the timber.

Associations

A **muffle** is an internal chamber in a small kiln, made from heat-proof (refractory) material, which protects materials from direct contact with flames and gases in the kiln.

See: **enamel, distortion, glaze, oxidation, pyrometry, season, shrinkage, thermodynamics, warp.**

Knot

Pronounced: NOT (*o as in hot*)

Origin

From Old English *cnotta* from *cnyttan* meaning *to knit, to tie*. It came to mean anything bunched, twisted or entangled and then the shape of the result of such an action.

Meaning

A knot is a hard-massed part of a tree where a branch shoots out. A live knot cannot be knocked out of the wood; a dead knot can be loosened and taken out. In a piece of prepared timber, it is a cross-grained lump, which ocassionally falls out of the piece to make a hole. It is also called a **knur** or a **gnarl**, which can also mean twisted and covered with lumps.

Associations

In painting, **knotting** refers to putting a solution of shellac in spirit on knots in wood to prevent them from exuding resin.

See: **grain, knurling, resin, paint.**

Knurling

Pronounced: NURL-ING (*u as in fur*)

Origin

From an old German word *knorre* meaning *a knot in a tree* or *a hard mass of something*. It came to mean a small knob of something or a small ridge.

Meaning

Knurling is a raised pattern of incised parallel or criss-cross lines on a piece of metal. The pattern is obtained by forcing a hardened, grooved roller or *knurl* into the metal workpiece as it revolves on a lathe. A parallel pattern is produced by a tool which has one roller; a criss-cross or diamond pattern is produced by a tool which has two rollers set one above the other. It produces one groove which is right-handed and another which is left-handed. Often cylindrical tools, such as a punch, are knurled so that they can be gripped easier but also as a decorative effect.

Associations

See: **groove, knot, lathe, mill, pattern.**

diamond knurling

straight knurling

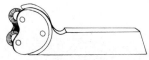

a knurling tool

Lacquer

Pronounced: LAK-A (*1st a as in cat, 2nd a as in ago*)

Origin

Lacquer has two different meanings and two origins. One is from the Hindustani word *lakh* meaning a resinous substance produced by the females of many species of insects, especially the *coccus lacca*. The insects feed on the sap of the Sumae tree in Siam and India. The resin hardens around them and their young. It protects them from predatory birds and other insects. This secretion forms into flakes and is collected at night. The word changed from the Portugese *lac* (from the name of the insect) to the French *lacre* (meaning *sealing wax*) and then to *lacquer*. The second meaning and origin is from the sap of the lac or lacquer tree which has grown in Eastern Asian countries for thousands of years and is a native of China. Lacquer ware was used in China in the 7th century B.C.. It was introduced into Japan in the 8th cenury and developed into one of the outstanding art forms in Japan in the 17th century. Lacquer work became very popular in Europe in the 18th century.

Meaning

Cellulose lacquer is a very hard, durable, heat-resistant, coloured varnish. When built up in layers, it is hard enough to be carved. There are natural lacquers (gum-lacs) made from the secretion of a Sumae tree and from the Lac tree, and man-made lacquers produced from plastic resins and acrylics. Natural lacquers tend to turn yellow and become brittle after long exposure to sunlight. Synthetic lacquers (usually made up of nitro-cellulose dissolved in an appropriate, volatile solvent, to which pigments, resins and plasticisers can be added) are easy to apply by spraying, weather well, dry very quickly by evaporation of their solvent and produce high-gloss finishes (but they can be matt or flat) which can be polished to a high lustre. They are used extensively, for example for the bodywork of automobiles and to prevent tarnishing and oxidation of polished metal surfaces. Cellulose lacquers are generally sold in a concentrated form and a thinner has to be added to them before they can be applied by brushing, dipping or spraying.

Plastic lacquers are made up of thermosetting resins (e.g. urea/melamine, urea/alkyd, polyester, etc), which, by chemical reactions, are cured (hardened) by the addition of a catalyst or accelerator. There is a very wide range of plastic lacquers, each of which has different properties. They are usually applied by a spray gun, but rollers and brushes can be used.

photograph by kind permission of Croda Paints

photograph by kind permission of Mirotone

Associations

Japanning refers to an 18th century word which describes artefacts of tin, wood or paper, which were painted black and then covered with lacquer to produce a high gloss sheen. The artefacts were relatively inexpensive imitations of expensive Japanese lacquer ware.
Shellac refers to *lac* from a Sumae tree which is melted and formed into thin plates or shells (*shell-lac*). It is now a type of varnish.
See furniture of the Louis XV period (1723 - 1774), when lacquer was much used.
See: **accelerator, acrylic, catalyst, enamel, finish, oxidation, pigment, plastic, resin, solvent, synthetic, varnish, viscosity.**

Laminate

Pronounced: LAM-IN-AT (*1st a as in lamb, i as in pin, 2nd a as in late*)

Origin

From the Latin *lamina* meaning *a thin scale, layer, sheet or flake* of metal, bone, rock, vegetable tissue, etc. and from *ate* a suffix forming a verb. That is *to layer*. The discovery of melamine (a hard clear resin plastic) in 1939 led to the production of laminated plastic sheets, in many colours and designs, which were used for many purposes in homes, industries and businesses.

laminating process

Meaning

A laminate is the sandwiching and bonding of two or more layers of paper, cardboard, plastic, wood, plywood, foil, glass-fibre, etc. by means of an adhesive or thermosetting resin. The layering of materials increases not only the thickness but also the strength of the laminate. Materials which are laminated are usually stronger than solid materials of the same weight and dimensions (e.g. layers of glass and fibre-glass for security windows). A laminating machine, a heated hydraulic press, is usually used in industry for the process, where sheets are pressed between heated plates (called **platens**) until the resin cures. Laminating in small workshops involves the softening of the material to be used (e.g by the use of a steam-chamber for timber) and the shaping on a **former** of the object.

a laminated dining chair

Objects which are laminated are often lipped (edged) with a veneer. Laminating is also used in book binding where a layer of plastic is placed on the the front and back of a book cover in order to protect it and to give it a glossy effect. Wood laminates in a variety of timbers (usually with the grain of the layers going in one direction, which makes the laminate very strong) are particularly effective where an article is needed which bends or curves. They are used for golf clubs, tennis and squash raquets, skis, roof beams (often with curved arches), chairs with laminated backs , boats, canoes, caravans and decorative wood utensils, etc..

Laminated plastic sheets are usually in veneer form

veneer from a former

rather than as boards. The veneers (layers of synthetic resin, often melamine, bonded under heat and pressure) are adhered to a base which is stable (so solid wood is not used as it can move with atmospheric changes), such as particleboard, plywood, hardboard or blockboard. The plastic laminates, which can have gloss or matt finishes, are available nowadays in a wide variety of designs, colours and patterns, which includes simulated textures and photographs. The plastic laminates are used for veneering furniture, for panelling and for working surfaces (e.g. in kitchens, bars and offices) as they are hard, durable and easy to keep clean and as they are resistant to most stains.

Associations

a laminate cutter

A **thermo-setting** process is where materials are subjected to heat and pressure. **Laminex** is a plastic laminate. **To Delaminate** is to split a laminated plastic material along its layers.

See: **adhesive, cramps, curing, finish, form, furniture, hardboard, hydraulic, lipping, particleboard, plywood, template, thermosetting, veneer.**

Lap

Pronounced: LAP (*a as in cat*)

Origin

From Middle English *wlappen* meaning *to wrap, surround, encircle.*

Meaning

A lap is a rotating disc covered with leather or some such material, which is covered with an abrasive substance and is used for cutting and polishing gems and metals.

To lap is to rub a surface so that material is rubbed away or so that it is polished.

Lapping is often used on a lathe as a means of finishing precision parts, such as the bore of a hardened bush. For instance, a lap, consisting of a material softer than the workpiece (e.g. copper, aluminium or cast iron), is covered with a suitable abrasive, made into a paste with oil, and it is moved back and forward through a bore.

In carpentry, **lap work** refers to a workpiece that has **lap-joints**; they are joints where one piece of wood overlaps another.

In metal work, **lap-rivetting** and **lap welding** both have metal parts which overlap before a riveting or welding process begins.

Lapping also refers to blemishes in painting or varnishing where the marks of brush strokes can be seen to overlap one another.

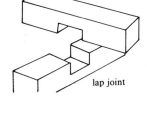

lap joint

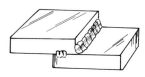

a tack-welded lap joint

Associations

See: **abrasive, bore, buffing, fillet, finish, grind, honing, joint, lathe, paint, rivet, rout, scarfing, seam, weld.**

Laser

Pronounced: LAZA (*1st a as in late, 2nd a as in ago*)

Origin

Laser is an acronym of *Light Amplification by Stimulated Emission of Radiation.* In 1917, Albert Einstein, a famous German mathematician and scientist, proposed the mechanics of stimulated emission - the principle of the laser action. A number of American scientists contributed to the development of the laser

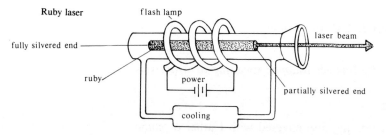

A ruby laser, which can produce for a few millionths of a second light ten million times more powerful than the light of the sun.

in the 1940's and 1950's but T. H. Maiman, who worked for the Hughes Aircraft Company in California, made the first laser in 1960 and J.Townes of Columbia University is generally credited with being the first person to apply laser knowledge to the making of practical things.

Meaning

A laser produces a narrow beam of light of only one wave-length (that is one colour) which goes in only one direction. The light is very highly concentrated (e.g. it can focus on a spot one millionth of a metre (a micrometre) or even less) and generates intense heat (ten million times the intensity of sunlight). It cuts very easily the hardest of substances (e.g. it can burn holes in steel plate and set carbon on fire). Yet, it can be quite easily controlled. It can be used for low-cost, high-speed engraving. The number of uses to which it can be put increases continuously and it will be used much more in the future in all kinds of metal and plastic work and in kinetic and optical art and design. There are several kinds of lasers, used for varying functions, but their basic principles are the same.

Lasers are part of what is called the **chipless technology**; that is the use of devices which can cut hard substances without producing chips of waste material. **Laser beam machining** is a method of cutting hard metals by melting them using an intense laser beam. Others in this category are **elecrical discharge machining** which uses electrical sparks to remove particles of metal from a workpiece, **ultra-sonic devices** which can disintegrate materials by high-speed vibrations and **plasma-arc torches** which produce ultra-high ionised gas which can melt and cut metal.

Associations

See: **engraving, holography, kinetic, sculpture, weld.**

Laths

Pronounced: LATH (*a as in far*)

Origin

From Old English *laett* meaning *a plank, strip of wood.*

Meaning

A lath is a piece of timber in a thin strip, usually about 2 to 3 cms. wide and thick. It was used formerly to create a trellis for the support of plaster for walls. Trellis work is used in gardens to support tall and rambling plants and to act as a *divider*.

Associations

See: **fretted**.

Lathe

Pronounced: LATH (*a as in late, th as in that*)

Origin

From Danish *lad* meaning *a structure* or *frame*. The Latin *tornus* mean *lathe*. See: **turn**.

The lathe almost certainly had its origin in the hunting bow. A bowstring was wrapped around a workpiece and the bow was moved quickly back and forward. The workpiece turned. A cutting tool was held against the workpiece, which was shaped as needed. The **bow lathe** was used in the Middle East before 3000 B.C.. A Greek, Theodore Samos, is credited with the introduction of the first turning machine, the **pole lathe**, about 600 B.C.. Wood turning on a pole lathe continued until the 14th century, when wheel-driven lathes and foot-treadle lathes were used. The first screw-cutting lathe, which was made of wood, was developed by a French man, Jacques Besson, in 1568.

Henry Maudslay, an Englishman, is credited with the invention of the first all metal screw-cutting lathe in 1800 (until then the machinery had been fitted to a wooden frame). An America, Thomas Blanchard, invented the automatic profile lathe in 1818. The invention in 1849 of a **turret lathe** (also called a **capstan lathe**) by three Americans, Frederick W Howe, Richard S. Lawrence and Henry D Stone, which allowed up to eight cutting tools to be rotated in sequence without

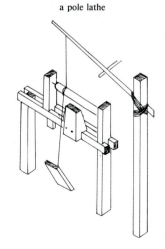

a pole lathe

stopping for tool changes, was a major advance in the mass production of metal components.

Meaning

A lathe is a stationary machine for shaping wood, metal ivory and other materials into circular, cylindrical or moulded shapes by rotating the material against a power-driven cutting tool. A variety of cutting tools can be used, which can be moved in several directions relative to the workpiece. The speed at which the workpiece revolves and the cutting tool operates can be changed according to the cutting required. Some lathes have attachments, such as a band saw, a circular saw, a planer, a sander etc., which extends their range of operations. A lathe can be used for contour turning, freehand turning, reaming, filing, boring, screwcutting shaping and polishing.

Associations

a **lathebed** is the lower framework of a lathe with slots from end to end for adjustment. The **lathehead** is the **headstock** (which holds the bearings of the revolving part of the machine) of the lathe. A **turret** is a special tool-post holding several different tools. The tools can be rotated to bring each one against a piece of work. **Parting off** is the process of paring wood or metal which is on a lathe. A **bench lathe** is a small lathe which can be mounted on a bench.

See: **arbor, boring, chisel, chuck, clearance, contour, face, indexing, jig, knurling, lap, machine, mandrel, paring, ream, screw, shape, turning.**

using a parting tool

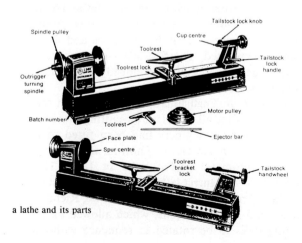

a lathe and its parts

148

Lens

Pronounced: LENZ (*e as in lend, z as in zebra*)

Origin

From the Latin *lens* meaning a *lentil*, which is the grain of a bean plant with a distinctive elongated-oval shape. Lens as we know it today derived from the shape of the lentil.

Roger Bacon, an English Franciscan monk who made a number of inventions, recorded the magnifying properties of a simple lens in the 13th century. Spectacle lenses were introduced into Europe in the mid 14th century and were commonplace by the 16th century. In 1590 a Dutch spectacle-maker, Zacharias Jansen, placed two lens together, one convex and the other concave, in alingment and was able to magnify an object. This is usually considered the invention of the microscope. An American, Benjamin Franklin, invented bi-focals in 1760. The first plastic lens was made in 1938.

Josef Max Petzral, a Hungarian, designed the first lens specifically for photographic use in 1840.

Meaning

A Lens is a piece of transparent substance (e.g. optical-quality glass or plastic) with one or both sides curved (either convex or concave) which is able to concentrate or disperse (refract) light rays in a camera, a telescope, spectacles or some similar instrument. Its purpose is to bring together the light rays reflected from any point of a **subject** onto the lens, at a point some distance away on the further side of the lens. The greater the convexity of the lens the shorter is this distance; that is the shorter is the focus. A lens is described as being of so many millimetres focal length, which means that parallel rays of light from a distant object are brought to focus on a plane this distance behind the lens. The opening in the lens (called the **aperture**) can be changed (by a mechanism called a **stop** or **diaphragm**) to control the amount of light which passes through it. The size of the opening is called a **setting** and each has an **f number**. A typical range of setting numbers is: f22, f16, f11, f8, f5, f4, f2.8. No lens is perfect but by the use of more than one type of glass (different glasses refract light differently) and the combination of lenses as a

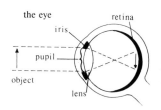
the eye

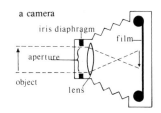
a camera

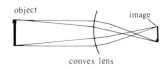
convex lens

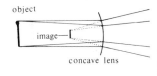
concave lens

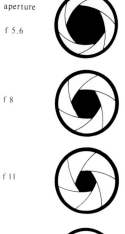

large aperture
f 5.6

f 8

f 11

f 22
small aperture

unit almost perfect results can be achieved. Today lenses are **anastigmats**. That is they prevent *astigmatism*, which is the inability to bring light of different colours to one focus. The angle of a lens determines the amount of a subject in a picture. There are special **wide angle** lenses (e.g. a **fisheye lens** which gives an angle of 180°) and **narrow angle** lenses (e.g. a **telephoto lens**), which have specialised functions.

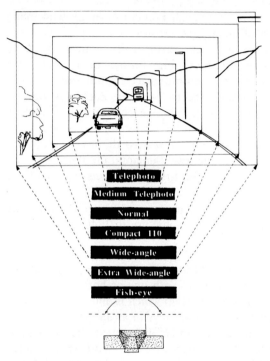

Associations

A **fresnel lens** is a thin, light lens which looks like a series of rings, each one of which is part of a normal lens. It is moulded from heat-resistant glass and is used in beam-focusing spotlights and rear-projection devices.
A **diffusion disc** is a device which is put on a camera lens to give a soft or slightly out of focus effect. It is often used for portraits.
See the pioneering work in lenses for photography of William Hyde Wollaston (1766 - 1828), Charles Louis Chevalier (1804 - 1859) and Josef Max Petzral (1807 - 1891).
See: **contrast, exposure, filter, focus, negative, scale shape, stop**.

Line

Pronounced: LIN (*i as in fine*)

Origin

From the Latin *linum* meaning *flax*, from which *rope* was made. Rope was used to mark out a series of objects in horizontal order. The word then came to mean a long, narrow mark made horizontally on a surface. Drawings using lines have been made since pre-historic times. The oldest use of line to communicate visually is that in the drawings of animals scratched on cave walls 30,000 years ago (See: **paint**).

Meaning

A line is the path of a visual moving point (e.g. from a pencil, pen, chalk, crayon or brush) as it draws across a surface. The artist Paul Klee (1879 - 1940) said:"*A line is a dot that has gone for a walk*". Its visual impact is usually created by its contrast with the surface on which it is drawn. Artists give much thought to their use of line, as different forms of line convey different ideas and feelings, as follows:

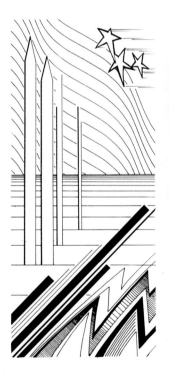

- **horizontal lines** suggest rest, repose (e.g. a fallen tree), but can also suggest movement and speed, particularly when there are a series of such lines (e.g. trailing lines from a comet or spaceship);
- **vertical lines** suggest alertness, life, dignity and nobility. The eyes are carried upwards (e.g. a church steeple, an *upright* man);
- **diagonal lines** suggest movement and action and jagged or zig-zag lines suggest movement, pain or tension;
- **curved lines** suggest delicacy, tranquility and, when free-flowing, gentleness, grace and sensuousness;
- **round lines** suggest smoothness, restfulness or completion.

Lines can also convey a sense of texture (by cross-hatching, stipple, etc.) and their weight (i.e. thick or thin lines) can suggest ideas. For example, a heavy line can emphasise importance or in some cases movement as the eye is attracted towards the emphasised line.

In graphic art, line refers to a half-tone dot frequency. That is a forty line screen is one that produces forty dots per *centimetre*.

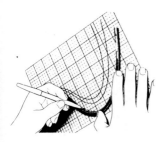

Associations

Linearity is the reliance of a painter or designer on lines for a drawing rather than colour or tone. A **vector** is a mathematical or geometric line denoting size and direction. **Lineal** is along the length of a material. Timber is measured in lineal metres. A **spline** is a flexible rule or rod which is used to draw curved lines.

See: **contour, cross hatching, drawing, elements, elevation, geometric, half-tone, mechanical drawing, scoring, scribing, shape, spline.**

Lintel

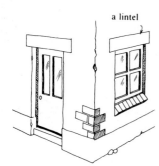

a lintel

Pronounced: LINTAL (*i as in bit, a as in ago*)

Origin

From the Old French *lintel* meaning *threshold*, the point of entry into a building.

Meaning

A lintel is a horizontal piece of timber, stone, steel or reinforced concrete spanning the top of a door or window or other similar opening. It supports walls and the ceiling or roof above the opening. An alternative term for a lintel is a **transom**.

Lipping

Pronounced: Lip-ing (*i as in pin*)

Origin

From the Old English *lippa* meaning *lip* and also *edge*.

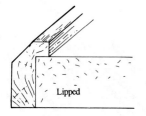

Lipped

Meaning

Lipping is the covering of an exposed edge of particle board, plastic laminate, veneers and manufactured boards with a thin layer of wood, plastic, or anodised metal or combinations of edges (mouldings) made from various materials. An alternative term is **edge stripping**.

Associations

See: **anodise, laminate, moulding, particleboard, veneer.**

Machine

Pronounced: MA-SHEEN (*a's as in ago, ee as in see*)

Origin

Machine is from the Latin *machina* meaning *an engine, contrivance, frame, device, trick*.

Some woodworking machines (e.g. jig saws, circular saws and reciprocating saws), which were operated by animal or man power or by water or wind, and crude wood turning lathes existed before the 1700s but most wooden objects before 1800 were produced using hand tools. The use of machines developed dramatically during the first three decades of the 19th century - the start of **mechanization**. Machines for working wood developed at the same time as machines which would work metal. A number of designers, engineers and inventors collaborated in the early 1800s (such as Samuel Bentham (English 1757 - 1831), Sir Marc Isambard Brunel (French 1767 - 1849), Joseph Bramah (English 1784 - 1814) and Henry Maudslay (English 1771 - 1831)). Hundreds of machines (such as planing, milling and shaping machines) for both wood work and metal work were invented during the 1800s and the early 1900s as industrialisation, importing and exporting, mass production and automation grew at an enormous speed. By the 1920s, industry had developed a "*machine style*", where products were designed so that they could be produced by machine techniques on a mass-produced scale. Machine tools had to be invented to produce the machines needed. For instance, it was only with the invention in 1775 by John Wilkinson of a precision boring machine that efficient steam engines could be produced which had large cylinders precise in interior size so that steam would not leak between a cylinder and its piston. The 20th century (which is sometimes called the *Second Industrial Revolution*) has seen the development and sophistication of the early machines and the introduction of automated systems for machine tool control, using numerical systems and computers. An American, Frank Stuelen, invented numerically-controlled machine tooling in 1952.

Meaning

A machine is a device that can replace human or animal effort to accomplish a physical task. A machine transmits or tranfers energy by motion through the **mechanism** of the machine which usually has several interconnecting parts, each of which has a specific function. Motion is transferred from one part of the mechanism to another in three basic ways: by a linkage, such as a crank and a connecting rod; by direct contact between gear teeth and a cam and follower; by a wrapping connector, such as a belt, rope or chain. Some machines have a combination of these methods.

Machinability describes the ease or difficulty of machining a material, which is measured by the **machinability index**. For example, steel is difficult to machine, magnesium alloy much less so. Magnesium alloy has a machinability index of 500 to 2000; tool steel has an M.I. of 34.

symbol to indicate a surface is to be machined

Associations

A **machine tool** is a mechanically-operated, adjustable tool which is used when working on materials such as metal, wood or plastic. **Molybdenum** is a very hard steel, which is often used for machine-cutting tools.

See the pioneering work of: Joseph Bramah (b. 1748), Henry Maudslay (1771 - 1831), Clement (1779 - 1844), Fox (1789 - 1859), Roberts (1789 - 1864), James Nasmyth (1808 - 18990), Joseph Whitworth (1803 - 1887).

See: **cam, coolant, cybernetics, fit, flute, indexing, lathe, particleboard, pawl, pinion, ratchet, reciprocate, relief, swarf, turn.**

Malleable

Pronounced: MAL-IABAL (*1st a as in cat, 2nd & 3rd a's as in ago, i as in ink*)

Origin

From the Latin *malleus* meaning a *hammer* and *malleare* meaning *to beat with a hammer.*

Meaning

A malleable object is one that can be hammered, rolled or pressed into a shape required without rupture or fracturing. Gold is the most malleable metal, followed (in order of degree of **malleability**) by silver, aluminium, copper, tin, platinum, lead, wrought iron and soft steel.

To malleate is to hammer or to form a material as flat

as a plate or as thin as a leaf by hammering. **Malleiform** means *hammer-shaped*.

Associations

A **mallet** is a hammer with a heavy wooden head.
See: **dapping, drawing, ductility, emboss, form, gilding, hammer, moulding, rivet, shape.**

Mandrel

Pronounced: MAN-DRAL (*1st a as in cat, 2nd a as in ago*)

Origin

From the French *mandrin* meaning an *axle, winch* or *crank*.

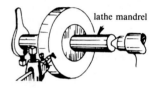
lathe mandrel

Meaning

A mandrel is a shaft of hardened, ground steel fitted to a lathe on which articles are fixed which cannot be turned by any other method, as they would be difficult to hold in a chuck. It is slightly tapered on the outside so that it can be pressed into the bore of a work-piece with a driving fit. Lathe mandrels are available in a variety of sizes and types (e.g. **the solid mandrel, the expansion mandrel** the **threaded mandrel** the **taper-shank mandrel** and **the gang mandrel**).

A mandrel is also the axle of a circular saw and a cylindrical rod around which metal or some other materials can be forged.

Mandrels are also tapered stakes of lustred hard steel, used for shaping and obtaining the size of rings or for making silver boxes, trays etc.

An **arbor** is an alternative term for a mandrel.

an expansion mandrel

Associations

An **axle** is the spindle upon which a wheel revolves and the connecting rod of the wheels of vehicles. A **crank** is the part of an axle or shaft bent at a right angle (as an arm or lever). Its function is to turn reciprocal motion into circular motion or vice versa.
See: **boring, fit, forge, lathe, reciprocate, turn.**

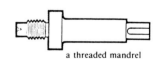
a threaded mandrel

Marquetry

Pronounced: MA-KET-RY (*a* as in *far*, *e* as in *let*, *y* as in *duty*)

Origin

From the French *marqueter* meaning *to variegate*. That is to mark with irregular patterns of different colours to produce varied effects. The art was practised in Ancient Egypt, Greece and Rome. The wood carvers of Italy developed the technique of wood **inlays** (called *intarsia*) to a very high level in the 15th and 16th centuries, often working from **cartoons** produced by eminent artists, such as Botticelli (1444-1512). The process became popular in Holland and Germany in the 17th century and became a distinctive feature of their furniture. Later it was highly regarded in France. It was first introduced into Britain during the middle of the 17th century.

An intarsia panel in the Church of Santa Maria Novella in Florence

marquetry

Meaning

Marquetry is a decorative woodwork process where shaped pieces of wood of the same thickness, or wood **veneers**, of different colours and grains are glued to a wood ground (in much the same manner as a jigsaw), in order to create a design or picture for furniture or wood utensils. Wood is the main material used but sometimes slivers of metal, tortoise-shell, mother-of-pearl, ivory or other textured materials are inlaid into the designs. The invention of a marquetry **donkey** (a type of jig saw) in 1780 allowed **marqueteurs** to cut up to ten pieces of veneer at a time, which increased production substantially. Floral and arabesque designs were popular in marquetry.

Associations

Inlays of wood are called **intarsia** in Italy. The inlay process is different from the marquetry process, in that intarsia fit inside a base on which they are placed but marquetry veneers are placed on top of the base. André-Charles Boulle (1642-1732, born in Paris of Swiss parents), a cabinet maker to Louis X1V of France, produced marquetry which was inlaid with ornamental patterns in brass, pewter, copper, enamel, or tortoise-shell. This kind of work is known as **Boulle work**. One of the finest **marqueteurs** was the German cabinet maker David Roentgen (1743-1807).

See: **furniture, inlay, texture, veneer.**

an intarsia donkey

Matt

Pronounced: MAT (*a as in cat*)

Origin

From the Persian *mat* meaning *dead* and from this anything which is not bright and alive is matt or dull.

Meaning

To matt is to give a pebbled or grained texture and a dull, unglossed finish to a surface, in contrast to a smooth, polished, gloss surface. On a wood or metal surface, this is called **matting**. Water-based paints produce a matt finish; oil-based paints usually produce a gloss finish. A matt finish can be given to a work by the inclusion of a matt medium, or varnish can be used with a matt finish.

Associations

See: **contrast, glaze, grain, texture, varnish.**

Mechanical Drawing

Pronounced: MEK-AN-I-KAL (*e as in let, a's as in ant, i as in pin*),
DRAW-ING (*aw as in saw*)

Origin

Mechanical is from the Greek *mekhane* meaning *a contrivance, a device* and the suffix *ic* meaning *in the form of*. Mechanical came to mean anything associated with machines or devices. See: **drawing**.

Meaning

Mechanical drawing (also called **engineering drawing**) is a precisely - controlled form of drawing to convey exact and detailed information in graphic form by the use of drawing instruments. The term is used to describe all kinds of engineering and architectural drafting and work involving projections. From a mechanical drawing, it should be possible to construct what has been designed. **Graphicacy** is the ability to communicate precisely through graphics.

drawings for the construction of a boat

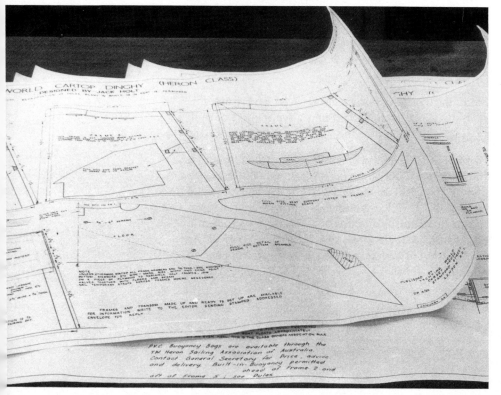

The stages of producing a mechanical drawing are: (1) drawing the borders and outlining the projections on the drawing paper; (2) constructing the projections faintly in pencil; (3) making the lines of the projections firm and clear; (4) inserting dimension and projection lines and putting in subtitles and notes; (5) drawing the title-block, parts-list or revision-table.

Some of the tools used to produce mechanical drawings are: bow pen, circle guides, divider-spacer, ellipsograph, French curves, lettering-guides, metal eraser shield, parallel rules, protractor, scalpel, set squares, snake curve, templates, T-square, try square and winged compass.

Associations

Dimensioning is the drawing of lines (straight or curved) on a drawing to show accurate measurements. **Projection lines** are thin lines which extend from the project outline but do not touch it. The lines show the ends of dimensions and extend a little beyond dimension lines. **Dimension lines** are thin lines, which sometimes have an arrow at each of their ends to indicate the length of the dimension. A dimension line ends when it meets a projection line.

A term which is gaining acceptance as an alternative term to mechanical and technical drawing is **technical graphics**.

See: **angle, cross hatching, cross section, draft, drawing, format, geometric, Golden Section, grid, Industrial Designer, line, perspective, plane, profile, projection, rectilinear, scale, shape, sketch, template, trammel.**

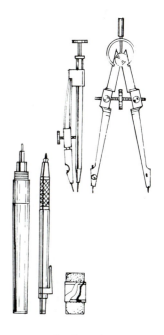

drawing tools

Medullary

Pronounced: MEDU-LARY (*e as in fed, u as in due, a as in ago, y as in duty*)

Origin

From the Latin *medulla* meaning the *marrow* or *pith* of plants.

Meaning

Medullary means consisting of or pertaining to the **medulla**, which is the centre (pith) of a tree. The **medullary rays** of a tree are the lines of cells in trunk of the tree which are shaped like the spokes in a bicycle wheel and radiate out to the cambrium layer.

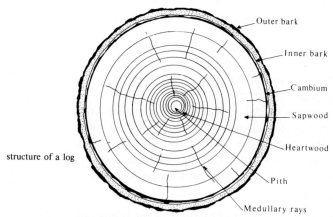

structure of a log

Associations

Exogenous is a botanical term used to describe plants, such as trees, which increase in size by adding rings of growth to existing concentric growth-rings beneath their bark.
See: **core, pored**.

Metallurgy

photomicrographs of the cystalline structure of metals.

The crystalline structure of brass under 500x magnification. The separate copper and zinc crystals are revealed.

Pronounced: META-LURJY (*e as in met, a as in ago, u as in fur, y as in duty*) OR METAL-URJY (*e as in met, a as in cat, u as in fur, y as in duty*)

Origin

From the Greek *metallon* meaning *metal* and *ergon* meaning *work*. A *metallourgos* in Ancient Greece was a *worker in metals*.

Metal was extracted from ores thousands of years ago. The Ancient Egyptian rulers 5500 years ago bathed in water which was conveyed by copper pipes from the river Nile to the baths in their palaces, so the study of metals has been practised for a very long time. Metallurgy as a profession, however, began during the Industrial Revolution in the 19th century when a demand for large quantities of iron and steel resulted in the construction of large iron works and intensive study of metals, especially ferrous metals. One of the most significant inventions in the 19th century was the Bessemer Converter in 1855, by an Englishman, Sir Henry Bessemer. It made possible the direct conversion of cast-iron into steel. The invention revolutionised steel manufacture. It greatly reduced the cost of the production of steel, which made it possible to use steel where previously iron had had to be used.

Meaning

Metallurgy is the science concerned with the study of the structure and properties of solid metals and alloys, of the extracting of metals from their ores and of the working and manufacture of metals. A **metallurgist** is a person skilled in metallurgy.

Copper zinc alloy 500x magnification.

Associations

Metallography is the study of the structural and physical properties of metals. The 20th century discovery of **X-ray diffraction**, which shows the minute structure of metals, has greatly increased our knowledge of the properties of metals.

Aluminium copper alloy 1500x magnification.

See: **alloys, fatigue, grain, noble metals.**

Mill

Pronounced: MIL (*i as in pin*)

Origin

From the Latin *mola* meaning *a place to grind things* from the verb *molere* meaning *to grind*.

In the second decade of the 19th century simple milling machines were being used. An American, Eli Whitney, built a simple milling machine sometime in the early 1820s, which cut grooves with a rotary cutter, although the concept had originated with an Englishman, Dr Hooke, in the 1660s. James Nasmyth, an Englishman, invented a small milling machine about 1830 to machine hexagonal nuts required for steam engines. These machines were improved on by George Bodner, a Swiss mechanic, in 1839. Further development was the invention in 1862 of the universal milling machine by an American engineer, Joseph R. Brown. The machine was able to cut flutes in twist drills used for cutting holes.

Meaning

A mill is a machine for grinding any substance by crushing it between two very hard surfaces.

To mill is to grind or stamp in a mill or to put ridges or furrows on any edge (as on the edges of some coins or adjusting screws).

Milled wood is timber which has been shaped by a **shaping machine** to produce a profile which is usually called a **moulding** (e.g. a skirting board, picture frame, weatherboard, floor board).

A **hand rolling mill** is a hand-operated device which is used to reduce the gauge of wire or the thickness of small pieces of sheetmetal.

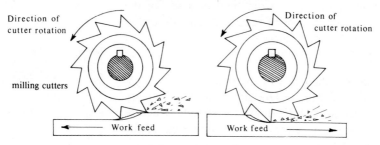

A **milling machine** is a power-driven, rotating cutting tool which has teeth called **milling cutters** of which there is a wide variety. It is used to cut gears and a variety of shapes, such as slots, grooves, shoulders of joints and dovetails. The word *mill* is also used for the cutters on a milling machine, such as a **cotter mill**, which is used for cutting slots and groves, and an **end mill** which is a cylindrical cutter which can cut at both its sides and end.

Associations

See: **arbor, flute, gauge, grind, groove, knurl, moulding, profile, shape.**

Mitre

Pronounced: MITA (*i as in file, a as in ago*)

Origin

From the Greek *mitra* meaning a *belt* or a *headband* and the Latin *mitra* meaning a *cap or turban*. A bishop's and archbishop's head-dress is called a mitre. Its distinctive shape where the top meets at about 45°, probably accounts for the use of the word for a kind of joint at an angle.

Meaning

When two pieces of material (e.g. of a moulding) meet at right angles, the *line* where they join, which bisects the 90° angle, is the mitre. The mitre is at 45°. A **mitre**

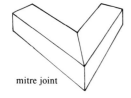

mitre joint

joint is a joint between two pieces of wood cut at an angle of 45°. A mitre joint is used at the corners of many wood constructions, such as picture frames, boxes, architraves and plinths. A **mitre block** and a **mitre box** are devices for holding a piece of material and a guide for a saw in the cutting of the material at an angle of 45°, prior to the making of a mitre joint.

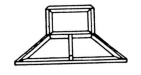

iron mitre template.

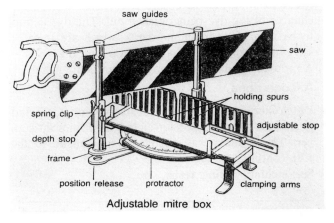

Adjustable mitre box

Associations

Mitre wheels are a pair of bevelled cog-wheels with teeth set at 45° and axes at right angles. A **mitre cramp** is a metal cramp which is used to hold a mitre joint while it is drying after being glued.

See: **angle, bevel, cramp, joint, key, plinth, saw.**

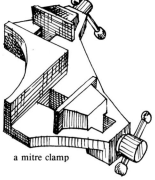

a mitre clamp

Mordant

Pronounced: MOR-DANT (*o as in more, a as in ago*)

Origin

From the Latin *mordere* meaning *to bite*. A mordant bites and holds fast to a material to become part of it. Mordants were used in India and Egypt more than five thousand years ago. It was originally thought that mordants *bit* open the surfaces of fibres to open passages through which dyes could get inside fibres.

Meaning

A mordant is an etching solution (e.g. nitric or hydrochloric acids or ferric chloride) which *eats away* those lines on a metal surface which are not protected by a **resist** material. The depth of the "*bite*" depends upon the strength of the mordant and also how long the metal is exposed to it.

In photography, it is used in **dye-toning** and some **colour-printing** processes.

In **gilding**, it is an adhesive size which holds gold leaf put on it.

In the **dyeing** of fabrics, it is a substance which causes fibres to swell and to absorb dye, which lessens the chances of the dye being washed out. The dye is said to be *fixed*. The main mordants, which are metallic salts, are alum, chrome, copper, iron, and tin.

A mordant is also a cleansing acid for some metals.

Associations

Dutch mordant is a solution of hydrochloric acid and potassium chlorate used for etching fine lines on copper. The formula was discovered in Holland in the 17th century.

See: **etching, gilding, resist, size.**

Mortise

Pronounced: MOR-TIS (*o as in bore, is as in miss*)

Origin

From Old French *mortoise*, which is probably from the Arabic *murtazz* meaning *fixed in, fastened*.

Two brothers in America called Greenlee invented the hollow chisel mortising machine about 1876.

Meaning

A mortise (or mortice) is a groove in wood designed to receive an appropriately-shaped end of another piece of wood to create a joint. The part shaped to fit exactly into the mortise cavity is called a **tenon**. **To mortise** is to join one thing into another.

Associations

A **mortiser** is a power machine used in the commmercial production of furniture which cuts a mortise by means of a continuous chain cutter and a square hollow chisel and auger. A **mortise lock** is a lock which is recessed into a mortise at the edge of a door.

See: **furniture, gauge, groove, haunch, joint, rout.**

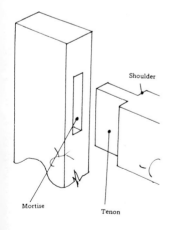

Motif

Pronounced: MO-TEEF (*o as in go, ee as in see*)

Origin

From the Latin *motivus* meaning *movement* or *to repeat a move*. Motifs in woodwork have been common throughout time. Many periods have distinctive motifs, such as the Tudor rose on furniture made during the period 1550 - 1600, or the curvilinear forms and floral motifs of the Queen Anne period in the 1700s.

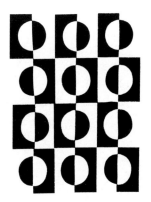

Meaning

A motif is a visual element, pattern, subject or theme that is repeated a sufficient number of times in the same or slightly different form in a composition or design to give it dominance. It is similar to a melody or theme in a musical composition. In sculpture, it refers to the manner in which a sculptor arranges figures he or she has produced.

Associations

See: **composition, design, elements, figure, pattern, replica, stencil.**

Moulding

Pronounced: MOL-DING (*o as in bold, i as in sing*)

Origin

From the Latin *modulus* which is the diminutive of *modus* meaning *measure, style, fashion, pattern, form*. Mouldings were used in Ancient Egypt, Greece, Rome and India. They were a distinctive feature of Gothic architecture, mainly in churches, in the 1200s.
The first commercial **injection moulding** machine was patented in 1926 in Germany. It became capable of full automatic production in 1937, and this revolutionised the production of cheap, mass-produced plastic commodities.

Meaning

A moulding is the product of a process where the faces and edges of wood or plastic are shaped to produce specific ornamental designs. Common moulding forms are: quarter round, half round, bead, cavetto and ovolo. Mouldings are used on picture and photograph frames,

Some types of wood mouldings

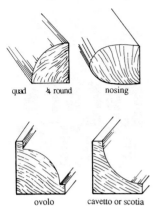

quad, ¼ round, nosing, ovolo, cavetto or scotia, bead, chamfer

picture rails, cornices, architraves, skirtings, and edges of some furniture. An **applied moulding** is a moulding which is shaped on a length of timber and then fixed to a work by nails, screws or an adhesive.

Moulding also describes the process of shaping any soft, malleable substance (e.g. thermoplastic) into a particular form or pattern.

Compression moulding is a process where thermosetting powder (e.g bakelite) is placed in a two-part mould (one part *male*, one *female*) and then subjected to heat and pressure. The softened plastic fills the mould and the moulded object is ejected when it is cured. Moulds, which should be strong in construction, are made from wood, metal, plastic, rubber, plaster, polyester, epoxy resins and glass.

Vacuuum forming is moulding by creating a vacuum on one side of a hot plastic material and using atmospheric pressure on the other side to push the plastic into shape. **Injection moulding** is a high-speed process used in industry where plastic granules are heated and the softened plastic is forced under pressure into a cool mould, where the plastic sets into a desired shape.

Associations

A **cornice** is a moulding where the ceiling joins the walls. A **moulding machine** is used for making wood mouldings. A **mould box** is a box in which molten metal is hydraulically compressed. **Mould facing** is fine powder or a wash (e.g an emulsion of silica flour, graphite and carbon) applied to the face of a moulding to ensure good definition of the mould and also to make the separation of the moulding from the mould as easy as possible. A **mould release agent** (e.g. wax or p.v.a.) is a substance which is used to coat a mould so that plastic material does not stick to the mould and make difficulties in releasing a moulding.

A **moulding plane** is used for forming mouldings. It is sometimes called a **match plane**.

See: **bead, blow moulding, casting, chamfer, cire perdu, clearance, contour, cure, die, draft, emboss, extrusion, face, fillet, form, lipping, malleable, mill, moulding, negative, pattern, plastic, plug, positive, profile, rout, shape, thermoplastic, thermosetting, vacuum forming, vulcanise, warp, web.**

Nails

Pronounced: NALZ (*a as in sale*)

Origin

From Old English *naegl* meaning a *finger nail* or *metal peg*. The Old English *naeglen* meant *to fasten with nails*. The earliest known nails were those found in a statue in Mesopotamia (now mainly Turkey), which dates back to 3500 B.C. In the Middle Ages (5th century to 16th century) nails were so expensive that carpenters rarely used them and made wooden joints to fix timbers. The first nail-making machines were made in the United States by Ezekiel Reed in 1786 and by Thomas Clifford in England in 1790.

In former times in England, the currency was pounds, shillings and pence with the symbols **£, s** and **d**. A hundred nails of a given size were sold for so many pennies ,e.g *4d* or *10d*. Today **4d, 5d**, and so on, indicate the nail length, with **2d** being small and **20d** large.

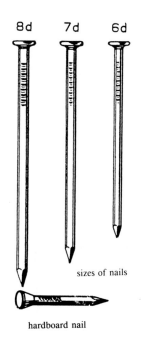

sizes of nails

Meaning

Nails are thin, pointed pieces of metal, usually with broadened heads, which are driven by a hammer or punch into a material (usually wood or a synthetic board) to fasten one material to another. They are usually made of cold drawn mild steel wire but where the nail comes into contact with water they are often made of other metals, such as copper, brass, aluminium, monel or silicon bronze. Most nails have diamond-shaped points but there are also chisel points, shear points, flat points and bifurcated (two-pronged) points. The nail point chosen depends on the type of holding power required and whether the timber splits easily or not. Some nails used externally are galvanized with zinc or plated with cadmium to prevent corrosion. Nails are generally described by the type of head they have (e.g. *diamond head, flat head, bullet head, spring head, clout head*) or by their use (e.g. *wall board nail, duplex nail, upholstery nail*) and by their size which is their length in millimetres, measured from the point to the top of the head, times the diameter of their shank or body in millimetres. Some small nails are called **pins** (e.g. panel pin, veneer pin, escutcheon pin). A **brad** is a long, thin, wire nail with a very small head.

Nails are made in a **nailery** by a **nailer**.

hardboard nail

Duplex nail

flex sheet nail

clout

flathead nail

particleboard nail

plasterboard nail

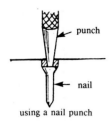
using a nail punch

Associations

A **nail set** or **nail punch** is a type of tapered, metal punch, used to set (*sink*) the head of a nail below the surface of timber.
See: **corrosion, fasteners, fixing, hammer, hardboard, panel, pinning.**

Negative

Pronounced: NEG-A-TIV (*e as in egg, a as in ago, i as in if*)

Origin

negative

positive

From the Latin *negare* meaning *to deny, say no, contradict* and *ivus* a suffix meaning *tending towards or having the nature of* (e.g. *descriptive* means *tending to describe*). It came to mean anything not positive or the reverse or opposite of something.

The negative-positive process in photography was invented by an Englishman, William Henry Fox Talbot in 1839.

During the early 18th century, Dufay (1699 - 1739) carried out experiments with glass and ebonite rods which had been electrically charged by rubbing them with silk. He propounded the first laws of **electrostatics** ("electricity at rest") which said that *like charges repel one another; unlike charges attract one another*. He stated that there are two kinds of charges, which we now call **positive** and **negative**.

Meaning

A negative is a photographic film on which lights and shadows of images are the reverse of what they are in reality. The light tones are reproduced as dark and dense, the dark tones as light and thin. It is the opposite of a positive. To produce a positive black and white picture, light is allowed to pass through the negative on to photographic paper, followed by developing and fixing.

Negative space refers to the areas in a composition or design between the outlines of shapes (objects or figures) and between shapes and the edges of the material being worked on. Shapes take up positive space. A composition or design results from the combination of negative as well as positive space.

A negative is also a mould which is shaped in the reverse form from the original prototype shape used.
In electricity, a point or electrode is said to be negative related to another point when it has a lower electric potential.

Associations

See: **anodise, bromide, develop, electrode, electronic, filter, gradation, grain, half-tone, lens, mould, positive, space.**

Noble Metals

Pronounced: NOBAL- METALZ (*o as in go, a's as in ago, e as in pet*)

Origin

Noble is from the Latin *noblis* meaning *well-known, famous, celebrated, superior*. It originated from *gnobilis* meaning *knowable*, which came from *gnoscere* meaning *to know*.

Meaning

The noble (that is *superior*) metals are gold, silver and platinum. They are different from most other metals in that they resist most chemical actions, corrosion and tarnishing in air or water and are not easily attacked by acids. Also, they keep their lustre when exposed to air. They are sometimes referred to as **perfect metals**.

Associations

See; **alloy, corrosion, gilding, metallurgy, oxidation.**

Notching

Pronounced: NOCH-ING (*o as in got, ch as in church*)

Origin

Probably from Old French *otch* meaning *to cut, to incise*. It was easier to say a notch, rather than an otch, so notch took over from otch.

Meaning

A notch is a nick or v-shaped incision or indentation in the edge or on the convex surface of a material.
It is also the cutting away of sheetmetal on a *develop-*

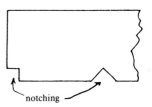

coping is similar in many respects to notching

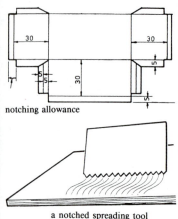

notching allowance

a notched spreading tool

ment to ensure that there is no overlapping or bulging of seams and edges when the development on the sheetmetal is formed to produce the desired shape. A **notcher** is a hand-operated machine which makes 90° cuts or notches in a metal workpiece. A **notching machine** produces blanks, using a die, which have been notched to make square pans and other ware.

In woodwork, notching is a method of joining timbers by **halving**, **scarfing** or **caulking** (driving metal plates together). Notching also describes the cutting out of a recess in one piece of wood so that it fits over another piece and checks that piece's movement.

Associations

See: **development, die, joint, scarfing, serration**

a hydraulic 90° notching machine

photograph courtesy of F I M, Bologna, Italy

Oxidation

Pronounced: OXI-DA-SHUN (*o as in box, i as in pin, a as in late, u as in sun*)

Origin

From the Greek *oxus* meaning *sharp* or *keen* and *genes* a suffix meaning *that which produces*. Oxygen, therefore, is that which produces keenness, and later vitality or life.

Oxygen, a colourless, tasteless, scentless gas, is essential to human, animal and vegetable life. It is also necessary for things to burn. *Oxidation* (or *oxidising*) is a 19th century word, derived from oxygen, which means *to cause to combine with oxygen*. The term *oxide* was coined by a French scientist G.de Morveau.

Meaning

In metal work, oxidisation refers to the production of oxides or metallic salts on metals (often referred to as **scale**) when they are heated or subjected to atmospheric conditions. **Corrosion** usually occurs on metals as they break down into oxides or metallic salts. Certain metals (e.g. aluminium, copper and copper alloys), however, require oxidation to prevent the metals from corroding. Developer for photographs oxidises and becomes useless when it is left in open trays.
Oxidisation is also the chemical reaction which occurs when oil paints, varnishes and enamels unite with oxygen fom the air, which results in their curing (hardening).

oxidising agent symbol.

Associations

Reduction is the opposite of Oxidation. Oxidation involves the loss of electrons; reduction involves the gaining of electrons.
See: **annealing, anodise, corrosion, curing, electronics, enamel, flux, kiln, lacquer, noble metal, paint, pickle, pigment, reduction, screw, solder.**

Paint

Pronounced: PANT (*1st a as in late*)

Origin

Paint of a very simple kind using natural pigments (such as clays, soot and chalk) was first used in cave paintings by the earliest hunters about 30,000 B.C. The wall paintings in caves at Altamira in Spain and those at Lascaux in France show that people thousands of years ago had knowledge of, and skills in, the use of, paint. Some maintain that the Dutch painter Jan Van Eyck was the first painter to use oil paint (using linseed oil). Whether it was he or not, it is certain that he and his brother used oil paint in their work in the 15th century (about 1400).
The most important development in modern paint technology came about 1918, when the first alkyl resins were manufactured from petroleum. In 1939, a German, Otto Bayer, produced the first polyurethane paint and Germany produced polyvinyl acetate (PVA) during the Second World War (1939 - 1945). The use of these

paint brushes

resins produced paint which had a very hard surface resistant to chemical attack. After the war, the first emulsion paints, which can be mixed with water, were produced. Recent years have seen increasing use by the paint industry of many types of synthetic resins (e.g. polymethyl methacrylate, epoxy resins, polyurethane resins).

Meaning

All paints comprise six ingredients: pigments, vehicles (or binders), solvents, and thinners, driers and surface active agents. Pigments provide colour and texture and affect the finish; vehicles (e.g. oils, resins and, varnishes) bind the other ingredients and affect the finish; solvents dissolve or disperes the vehicle; thinners dilute vehicles (the most common is mineral turpentine); driers accelerate drying time or solidification of the paint; surface active agents (such as detergents) wet pigments, which develops colour and improves both the flow of the paint and its adhesion qualities.

Paints were formerly used, and still are used chiefly, for their decorative effects but since the 17th century they have been used also as an important anti-corrosion agent. Paints are available to provide a full gloss, "eggshell" or matt finish.

a general-purpose paint spray gun

Associations

An **extender** is a substance which is added to paint to give it more body and greater adhesion or opacity. The first coat of paint which forms the base of other coats is called the **ground**.

See: **accelerator, catalyst, corrosion, filler, finish, glaze, hygroscopic, knot, lap, oxidation, pigment, preservation, priming, resin, solvent, vehicle, viscosity.**

Panel

Pronounced: PANAL (*1st a as in mat, 2nd a as in ago*)

Origin

From The Latin *pannulus* which is the diminutive of *pannus* meaning *a piece of cloth*, which was usually rectangular or square.

Meaning

A panel is usually a rectangular or square piece of material with raised margins or with a surrounding frame. It is frequently the distinct part of a finished surface, especially of doors and wainscots. **Panelling** is the dressing or decoration of doors (e.g cabinets) or interior walls with panels. A **panel strip** is a narrow piece of wood or metal used for covering the line or gap between two panels. A **panel pin** is a thin nail with a small head used in panelling.

A **panel beater** is a person who beats out metal panels for use on machines, such as a motorcycle or protective guards for power machines.

In electricity, a panel is a sheet of marble, slate or other material on which electrical devices such as switches and relays are mounted. It is also called a **switchboard panel** or a **control panel**.

Associations

A **panel saw** is a hand saw used in panelling, which has seven teeth every 2.5 cms.

See: **dress, finish, nails, particleboard, saw.**

Pantograph

Pronounced: PANT-A-GRAF (*1st a as in mat, 2nd a as in ago, 3rd a as in raft*)

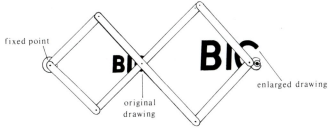

Origin

From the Greek *pantos* meaning *all* and *graphe* meaning *writing* or *drawing*. That is a device which is able to draw (copy) all that is in front of it. It was invented by a German, Christopher Scheiner, in 1603.

Meaning

A pantograph is an instrument for making copies of drawings on a different scale from the original drawing; that is for enlarging or reducing a drawing. It can also be used to distort the proportions of a picture or design. It is available in different standards from a simple, inexpensive toy to a high-grade, professional precision instrument. It has four hinged sections with a tracing point at one end and a pencil at the other. The hinges can be adjusted to control the scale of the copy.

Associations

See: **drawing, plane, replica, squaring up.**

Parallax

Pronounced: PARA-LAX (*1st and 3rd a as in tap, 2nd a as in ago*)

Origin

From the Greek *parallaxis* meaning a *change, alteration.*

Meaning

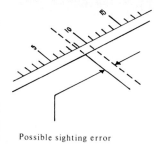

Possible sighting error

The parallax is the angle between two straight lines drawn to an object from different points of view. If one changes one's position when looking at an object at an angle, the object appears to have changed its position. It is an optical illusion which can lead to errors in making measurements. The parallax factor should be taken into account when looking down on an object to take measurements of it or when reading from a rule or gauge. When reading the scale on an electrical instrument, for example, the eye should be in correct alignment with the instrument needle or parallax can occur.

Associations

See: **angle.**

Paring

Pronounced: PAR-ING (*a as in care*)

Origin

From the Latin *parare* meaning *to prepare* and then *to trim* by cutting away parts not needed.

Meaning

Paring is the process of cutting thin shavings of wood from a wood object with a chisel or some other cutting tool, in order to shape a workpiece to a form required. A **paring chisel** is a tool with a thin blade, which is used for finishing off by hand a workpiece in wood. It is not struck with a mallet.

Associations

See: **dress, finish, lathe, shape**.

Paring a corner

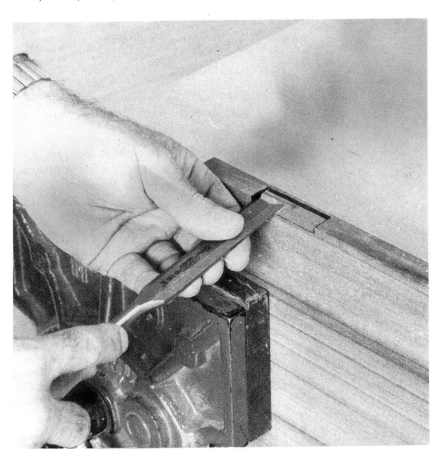

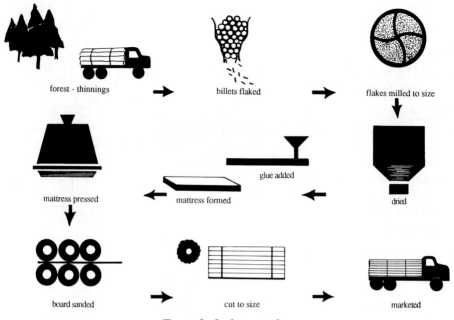

the manufacture of particleboard

Particleboard

Pronounced: PA-TIKAL (*first a as in far, i as in ink, 2nd a as in ago*), BAWD (*aw as in law*)

Origin

Particle is from the Latin particula meaning a *small part. Board* is from the Old English *bord* meaning a *plank.* The term was coined in the early 1930 to describe the boards of material produced from particles of compressed wood or other substances.

Particleboard was first produced on an experimental basis in Switzerland in the late 1930s but the first commercial production was in Bremen, Germany, in 1941. By the 1950s, particleboard had become established as one of the best boards to be used in the manufacture of inexpensive furniture and panelling.

Meaning

At an international conference held under the auspices of the Food and Agricultural Organization of the United Nations in 1957, particleboard was defined as: "*a sheet material manufactured from small pieces of wood or other ligno-cellulosic materials (e.g. chips, flakes, splinters, strands, shives, etc.) agglomerated by use of an organic binder together with one of the following agents:*

heat, pressure, moisture, a catalyst, etc.".
Ligno means *related to wood* and *cellulosic* means *related to* **cellulose**, which is the carbohydrate forming the main part of plant cells.

Particleboard is dimensionally-stable, free from defects (such as knots), can be readily produced in large sheets, is relatively light, is easy to work and can be machined in any direction and is relatively inexpensive. Consequently, it is now widely produced and used throughout the world, especially for wall panelling, ceilings, shelving, flooring, built-in furniture and units, radio and television cabinets, concrete forming and many types of inexpensive furniture.

A **hot press** is usually used to process the particleboard. The particle and sythetic resin mix are placed between two metal plates, called *platens*, which are heated by water, steam or elecricity and subjected to pressure. The resin is cured and hardens within the mix, which is dried and compressed into boards. This production method is called the *sandwich layer process*.

Associations

Bagasse is a fibre obtained from sugar cane, which is used in the manufacture of particleboard. Particleboard is sometimes called **chipboard, chipcore, chipcraft, shavingboard, flakeboard** and **wood-waste board**.
See: **catalyst, curing, furniture, key, laminate, lipping, machine.**

Pattern

Pronounced: PAT-AN (*1st a as in cat, 2nd a as in ago*)

Origin

From the Medieval English word *patron* meaning *master* (one who protects). The meaning has changed and now means the master or influential design which stands out in a work by its regular occurrence. Patterns have been used in design from the earliest times. Geometric patterns were made on potteryware in China 6000 years ago.

Meaning

Pattern refers to the visual forms or motifs which appear regularly in a composition, design or environ-

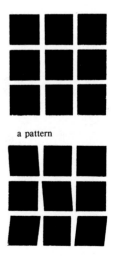

a pattern

anomaly in a pattern

ment in a systematic manner. Also, when elements such as colour, texture, size, volume, direction, shape, position, space are repeated, a pattern is created.

A pattern is the form in wood, plaster or styrofoam (in one piece or in sections) which is used as a prototype to make a mould in sand in a **sand casting** process.It is usually tapered to aid withdrawal. Molten metal is cast into the formed cavity left in the sand when the pattern is removed. **Pattern making** is the trade or skill of making patterns for use in sand casting.

Associations

Granulation refers to a pattern of tiny balls as decoration on metal.

See: **casting, clearance composition, design, drawing, extrusion, figure, filigree, gradation, hardboard, jig, knurling, motif, moulding, profile, replica, stamp, stencil, swage, symmetry.**

Pawl

Pronounced: PAWL (*aw as in saw*)

Origin

From the French *pal* meaning *a stake*. It is similar to a *pale* or to a *paling*.

Meaning

A pawl is a bar of metal (a lever) which is held against the teeth of a gear-wheel to prevent the wheel from moving in one direction, usually in reverse motion. It is a *catch* or *click*, which is usually controlled by a spring. It was commonly used to describe the metal

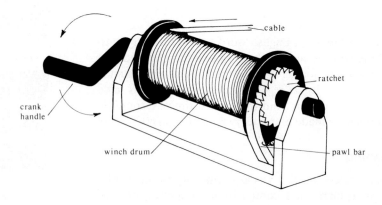

bar which prevented a capstan from recoiling. It is used with a **ratchet**, which is a gear with triangular-shaped teeth.

Associations

See; **machine, ratchet**.

Pene

Also spelled *pein, peen, pane* and *pean*

Pronounced: PEEN (*ee as in seen*)

Origin

From the Latin *pinna* meaning a *point*

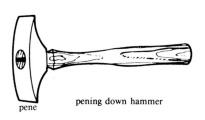

pene   pening down hammer

Meaning

The pene is the thin edge of a hammer head, which can have various shapes depending upon the type of indenting from a hammer one wants. It is opposite to the **face** of a hammer head.

To pene means to flatten out the end of a rivet with a hammer or to stretch and shape metal by hammering it, usually with a hammer which has a ball-pene end but also mechanically by using a machine.

Associations

A **pened down joint** is a joint used on sheetmetal to join the base to the body of a cylindrical or conical object.

See: **form, hammer, rivet, shape**.

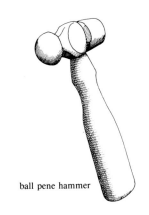

ball pene hammer

Perspective

Pronounced: PUR-SPEK-TIV (*u as in fur, e as in neck, i as in ink*)

Origin

From the Latin *per* meaning *by means of* and *spectare* meaning *to look* (Think of *spectator*.) and also *perspectum* meaning *to look or see through*.

The term was first used in the 5th century B.C. by the Athenian painter Agatharcus. Its principles were explained by the Greek mathematician Euclid. The basic rules of perspective, however, were first stated during the Renaissance by Italian artists Filippo Brunelleschi (1377-1446), Leone Battista Alberti (1404-1472), Piero della Francesca (1410-1492) and by Paulo Uccello (1397-1475), who is usually recognised as the author of the formula explaining perspective. The theory and the rules propounded have been developed over the years.

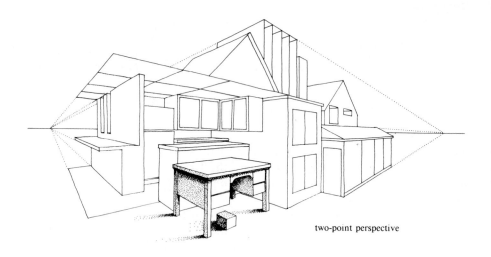

two-point perspective

one-point perspective

Meaning

Perspective is a man-made concept and technique by which three-dimensional objects or space can be visually represented on a flat (or nearly flat, i.e. a relief carving) surface. The aim of perspective is to create the **illusion** that the objects or people drawn have the same relative positions and sizes as the objects or people have in reality - when viewed from one specific point. To get the three-dimensional illusion, one must first draw a series of parallel lines which converge at a single point on the horizon (called the **vanishing point**). Between the lines, objects or people are drawn that gradually diminish in size. The key to perspective is this decreasing in size, which creates the illusion of **depth** and a three-dimensional effect. Early systems of perspective had only a single, central vanishing point (sometimes called **one point perspective**); later systems had two or more vanishing points, in order to attempt to produce greater naturalism.**Linear perspective** describes the system of parallel lines given above. **Aerial perspective** (first used by Leonardo da Vinci), which is also called **atmospheric perspective**, is a system to create the impression of distance and depth by gradually decreasing the clarity and colour brightness and contrast of objects and people, especially making them a hazy blue, as distance from the spectator seems to increase. Mountains in the distance, for example, should appear bluish in a painting. Leonardo da Vinci (1452-1519) advised painters to make the nearest building distinct and its real colour but the most distant buildings much less distinct and bluer. **Textural perspective** creates the illusion of distance and depth by gradually changing the textural appearance of things from sharply defined to hazy and dense. For example, the rough individual bricks in a house near the observer should be seen clearly with their mottled colours; bricks in a house in the distance should fuse into a smooth, one-colour wall.

Associations

Scenography is the art of drawing in perspective. **Sterography** is the art of representing the form of solids on a plane surface.

See: **drawing, focus, foreshortening, geometric, mechanical drawing, plane, space.**

Photogram

Pronounced : FOTO-GRAM (*o's as in go, a as in lamb*)

Origin

From the Greek *photos* meaning *light* and *gramma* meaning *things written or recorded*.

In 1803, two English men, Thomas Wedgwood and the chemist Humphry Davis, produced images of leaves and insect wings by placing them directly on sensitised paper and then exposing them to sunlight. The images, however, were not fixed and disappeared. In 1839, William Henry Fox Talbot, an Englishman, who invented the negative-positive process of photography, placed a piece of lace on a sheet of light-sensitive paper and produced the first *photogram*. Later artists, among whom were Man Ray (American 1890 - 1976) and Lazlo Maholy-Nagy (Hungarian 1895 - 1946), refined

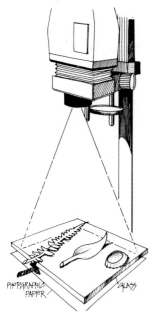

exposing a photogram from an object

and developed the camera-less photographic process. Lazlo Maholy-Nagy coined the term *photogram* for the process. They were very popular as an art form in the 1920's under the influence of the **Dadaist movement**.

Meaning

A photogram is the production of an image, without the use of a camera, by placing objects or making illustrations on transparent paper and then placing the paper on light-sensitive paper in a dark room and exposing it to a beam of light (e.g. from an enlarger, a desk-light or a torch). The photographic paper is then developed and printed. Also, *printing-out paper* (also called *studio-proof paper*) which has a low sensitivity to light, may be used for a design in subdued light and then it can be taken out into sunlight for exposure.

It has been called the purest form of photography or "writing with light". An infinite variety of effects can be achieved by varying the angles and distances of light (the objects create their own shadows which appear as permanent images after processing) and by moving objects during exposure.

Associations

See: **bromide, exposure, negative**.

Pickle

Pronounced: PI-KAL (*i as in pin, a as in ago*)

Origin

From the Old English *pekille* or Old Dutch *pekel* meaning *a salt (brine) or vinegar liquid* for preserving meat, fish, vegetables, etc..

Meaning

Pickle refers to an acid solution used in all kinds of metal work, including precious metals. It usually comprises either ten parts of water and one part of sulphuric acid or eight parts of water and one part of nitric acid. It is used to clean metals after an annealing process, before vitreous enamelling to ensure a secure bond between enamel and metal, to clean oxide or flux from silver, to prepare metals for soldering and for general metal cleaning where rust or mill scale has to be removed.

Associations

See: **annealing, enamel, finish, flux, oxidation, solder, vitreous.**

Pigment

Pronounced: PIG-MANT (*i as in pig, a as in ago*)

Origin

From the Latin *pigmentum* meaning *paint*. The first pigments were made from natural things, such as clays, soot, rocks, insects and plants. The first synthetic (man-made) pigments were developed by an English Chemist, William Perkin, who discovered the dye **mauveine** in 1856.

Meaning

Light is the source of all colour and pigments are substances which reflect, absorb or transmit colour. There are pigment molecules in everything (except transparent substances, such as water, air or pure diamonds) and they vary in form enormously, so some absorb or reflect a certain number of colours of the spectrum, some another number of colours. Daffodils, for example, have in them a pigment of yellow which they reflect and transmit to us.

Pigments are colouring agents in liquid or powder form which can be added to paints, dyes and liquid plastics. They are made from earths, animal and plant tissues or chemicals (metallic oxides). They are used to colour wood fillers, paints, plastics and fibre glass. As well as providing colour, pigments can affect the viscosity and drying time of a paint or enamel, can control the finish (e.g. gloss or matt) and some can also have protective properties.

The term **base** is sometimes used for a pigment.

Associations

See: **colour, filler, finish, lacquer, oxidation, paint, preservation, spectrum, vehicle.**

Pinion

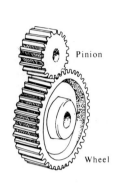

Pinion

Wheel

Pronounced: PINI-AN *i's as in ink, a as in ago*)

Origin

Probably from the Latin *pinna* meaning *a pin* or *a battlement* or from the Latin *pinea* meaning a *pine cone*. The terms pinion derives from the shape of a castle battlement with its indented wall-structure or the shape of an open cone with its indentations or gaps, like those in a gear-wheel.

Meaning

A pinion is a small-toothed gear or spindle which engages (or *meshes*) with another larger gear (called a **wheel**) as part of a **gear train**. When the gears are meshed, the two gears rotate in opposite directions and transmit rotary motion and force within a mechanism.

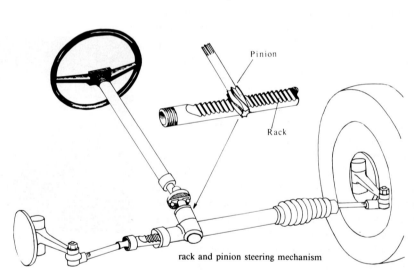

rack and pinion steering mechanism

Associations

A **rack and pinion** mechanism is used as a means of changing rotary motion into linear motion and vice versa. This mechanism is used, for example, in the steering mechanism of some cars and on the carriages of some lathes.

See: **machine, ratchet**.

Pinning

Pronounced: PIN-ING (*i as in tin*)

Origin

Pin is from Old English *pinn* meaning a *peg*, a thin piece of wood, which came from the Latin *penna* meaning a *feather* and later *a thin piece of wood or metal*. Today, a *feather* is used by carpenters to refer to a thin piece of wood which is used to tighten two pieces of joining wood.

Meaning

Pinning is a method of fixing timber using thin nails. The term refers also to particles of metal which clog the teeth on files, causing deep scratches.

Associations

See: **abrasive, fixing, joint, nails**.

Pitch

Pronounced: PICH (*i as in ink, Ch as in church*)

Origin

The origin of the term is obscure. It probably originated with the Old English *pician* meaning *to throw*, and the word took on many meanings,

Meaning

Pitch describes the distance between the centre of two teeth in a wheel or saw or between the point of one thread of a screw to the point on the next thread measured parallel to the axis. Pitch also means the number of threads per millimetre on a screw. **Axial pitch** (usually called just *pitch*) is the length advanced for each revolution of a screw thread.

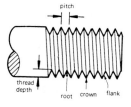

Fine pitch describes a screw with a relatively high number of threads per millimetre or a gear with unusually-small teeth.

Pitch also means the steepness of slope of a roof or other surface. Most houses have **pitched roofs**, which means they are not flat.

Pitch is also a dark brown or black resinous substance made from tar, which is used as base when hammering during some metal work.

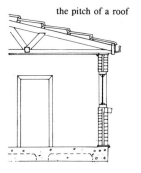

the pitch of a roof

Associations

See: **angle, cant, gauge, hammer, repoussé, saw, screw, thread.**

Pixelated Images

Pronounced: PIX-I-LA-TAD (*i's as in ink, 1st a as in late, 2nd a as in ago*), IMA-JIS (*i as in ink, a as in ago, i as in is*)

Origin

Pixelated derives from *pixels* which comes from two words, namely *picture elements*, where the common abbreviation of *pix* is used for *picture*.

a pixelated image

Meaning

Pixelated images (*pixelisation*) are produced by a computer programmed to create graphic material. The process involves the breaking down of an image, or small part of an image, into tiny areas called **pixels** (picture elements), where each pixel can be made one of many hues. A pixel can be 0.1 millimetres, or less, in length. That is 100 pixels to a square millimetre. The images on a screen are abstract in that they comprise thousands of minute squares. The technique is invaluable in restoration work of valuable paintings where small areas of paint have to be analysed. The more pixels there are on a visual display unit, the higher the degree of clarity and **resolution**. Computer graphic displays range, at present, from 320 by 200 pixels (which is a low resolution) up to 1024 by 768 pixels. The number of pixels that can be displayed on a horizontal line or a vertical column is called the **raster count**.

Associations

Pixelated is not to be confused with **pixilated**, as the word spelled with an *i* means crazy or drunk!
See: **computer graphics, electronics, geometric**.

Plane

Pronounced: PLAN (*a as in fame*)

Origin

From the Latin *planus* meaning *a flat and level surface*. In Ancient Roman times, carpenters had planes not unlike those today, with a wedge of cutting iron and a holding stock which had a clearance for shavings to escape. The first metal plane was invented by Leonard Bailey of Boston in the United States in the 1850s. An Englishman, Joseph Bramah (1748 - 1814), invented a planing machine in 1802 and another Englishman, William Woodworth, invented and patented a feed roller, rotary planing machine in 1828.

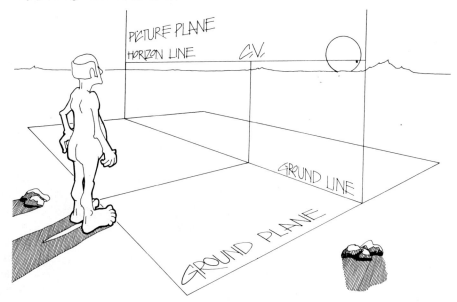

planing an end grain

Meaning

A plane is a two-dimensional shape on the surface of a material which gives no indication of volume, depth or projection but which when combined with other shapes placed beyond the plane (particularly when overlapping occurs) can give the illusion of depth and three-dimensional space. Planes are part of the construction of a composition and are not so much seen as felt. The planes which give direction can be in various positions in a composition, such as horizontal, vertical, diagonal etc.. **To plane** means *to make smooth and even* and a **plane** is any *paring and shaving cutting* tool used for producing level or smooth surfaces by removing thin shavings of wood. A plane has two main parts: a blade of sharpened steel, called **the cutting iron**, which is fitted into a wood or metal body piece, called the **stock**. There are three types of planes: **bench planes**, which are used on flat surfaces (e.g. *smoothing plane, jack plane* and *joint* or *trying plane* and *block plane*), **curve cutting planes**, which are used on curved surfaces (e.g. *spokeshave* and *compass plane*) and **special purpose planes**, which are used for mouldings, tongues, grooves and rebates (e.g. *router, rebate plane, bullnose rebate plane, rebate and fillister plane, side rebate* or *side fillister plane* and *combination plane*).

A **surface planer** is a machine which can plane surfaces and edges of wood at high speed and with fine accuracy. It can also cut chamfers, bevels, rebates and tapers.

Associations

A **spirit level** is used to obtain true vertical and horizontal planes during a process of construction. **Planary** means related to a plane. To **planish** is to make something smooth and to polish it. For this a **planisher** is used, which is a thin, flat-ended tool, often used for smoothing tin plate and brass work. A **planisher** is also a person who prepares copper plates for engravers. A **planometer** is a plane surface used in machine-making as a gauge for a plane surface. A **planigraph** is an instrument for reducing or enlarging drawings (as is a **pantograph**).

See: **bead, bit, composition, drawing, dress, finish, foreshortening, geometric, hammer, hardboard, jack, perspective, projection, rebate, rout, space, stop, thicknesser.**

Plastic

Pronounced: PLASTIK (*a as in pan, i as in stick*)

Origin

Plastic is from the Greek *plastikos* meaning *that can be moulded or shaped* from *plassein* meaning *to mould or shape*.

Natural plastics (which can be heated and shaped), such as amber, bitumen, lacquer resin, shellac, rubber and horn, have existed and have been used since before the times of the Ancient Egyptian, Greek and Roman and early Chinese civilizations. It is only since the early years of the 19th century that semi-synthetic materials (combining natural materials and man-made chemicals) and totally-synthetic materials (made entirely from man-made materials) have been invented and manufactured.

A great deal of research took place in the production of semi-synthetic and synthetic materials in the first few decades of the 19th century, but plastics were not manufactured on an industrial scale until the 1920s and 1930s. The first commercially-successful plastic material was *celluloid*, produced by a New York printer, John Wesley Hyatt, in 1870 (following work by a Brit-

189

a few of the wide variety of things made from plastics

ish chemist, Alexander Parkes, in 1862). Celluloid was made from a natural material (cellulose) combined with a synthetic, man-made material. In 1904, Leo Henrik Baekeland, a Belgian scientist living in New York, patented *bakelite* (phenolic resin or phenol formaldehyde), the first thermo-setting, heat-proof plastic made entirely from man-made substances. *Cellophane* was invented by Jacques E. Brandenberger, a Swiss chemist, in 1912. Based on work by the German chemist Hermann Staudinger (famous for his theoretical explanations of **polymerisation** and the nature of plastic materials), polystyrene, one of the most important plastics invented this century, was produced in Germany in the early 1930s. Nylon (with a chemical name of *polymide*), another early plastic which is used in many areas of life (e.g. from cloth to gear wheels to nuts and bolts), was invented by an American, Wallace Carother (and his associates), in 1934 and manufactured commercially in 1937. Since the 1930s, the number of different plastic materials, many with specific functions, has increased enormously and nowadays with the development of lamination and irradiation techniques, they have in many areas superseded metals. It is estimated that more than 50 million tons of plastics are used annually in the world.

Sources of Raw Material

Type of plastic material	Origin of intermediates
Polystyrene	Petroleum
P.V.C	Petroleum coal
Nylon	Coal
Cellulosics	Wood cotton
Phenolics	Coal
Polymethacrylate	Petroleum
Urea and melamine	Coal
Polyesters	Coal and petroleum

Meaning

Plastic describes a material (such as clay) which can be pressed into various shapes. The plural word, *plastics*, is used in the plastics industry for plastic products and plastic materials.

Plastic materials (with the exception of a few which are based on the chemistry of silicones) are synthetic resins based on the chemistry of carbon and are derived from coal tar and petroleum oil. They are made from chains of molecules which are as much as 10,000 times longer than natural molecules. They are shaped like huge straggling weeds with parts going in all directions. When these molecules are layered, they intertwine and become difficult to separate and are consequently very strong. When heat is applied to them they bond into one solid molecule which is different from the original molecules and far more complex. This process is called **polymerization** as **monomers** (meaning *one part*) change into **polymers** (meaning *many parts*).

Plastics is a term used nowadays for any non-metallic synthetic material (e.g. acrylic, polyester, p.v.c. polystyrene etc.) which can be moulded into virtually any shape. The mouldable material is said to have **plasticity**. They are notable for their strength, lightness, flexibility, inertness and their heat-resistant and electrical-insulation properties.

Associations

Plastic memory is the ability of *some* plastics to return to their original pre-moulded form after re-heating. This is also known as **elastic memory**. **Plastisol** is a liquid form of plastic which is used in moulding and is also used as a smooth surface-coating for some substances. A **plasticizer** is an additive to make plastics softer, more pliable and easier to mould into shapes required. **Ultra-violet** light from the sun over a period can cause damage to plastic materials. The addition of a **stabilizer** can prevent such harm. **Ceramoplastics**, which was developed in the 1960s, are heat-resistant plastics made by combining synthetic mica and glass.

See: **blow moulding, buffing, catalyst, cladding, cross-linking, curing, extrusion, form, glaze, grain, lacquer, resin, sculpture, sealer, solvent, synthetic, thermoplastic, thermosetting, vacuum forming.**

Plinth

Pronounced: PLINTH (*i as in ink*)

Origin

From the Greek *plinthos* meaning a *brick* or *square stone*.

Meaning

A plinth is the square base of a column or pedestal. It is also the projecting band at the bottom of a wall, which is usually called a **skirting** or **skirting board**.
In cabinet making, plinth refers to the square base section of a cabinet which raises the cabinet off the ground. It provides what is know as a **kick rail** or **toe rail**.

a plinth

Associations

See: **chamfer, furniture, grind, mitre.**

Plug

Pronounced: PLUG (*u as in dug*)

Origin

From an Old Dutch word *plugge* meaning *a piece of solid material which fits tightly into a hole*.

Meaning

Wall plugs are used in brick, concrete and masonary walls to hold nails and screws which are holding a wood or metal attachment to a wall. Without a tight-fitting plug in the wall a nail or screw would not hold in a wall. The plugs can be made of wood or plastic or be one of a number of patented metal plugs. A hole to take the plug is drilled into a wall and the plug is then driven into the hole. The plug cannot move and forms a very solid base for any nail or screw put into it.

a tampin

a dryvin

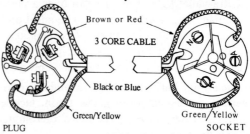

PLUG — SOCKET
Brown or Red
3 CORE CABLE
Black or Blue
Green/Yellow
Green/Yellow

192

A plug is also a device consisting of metal pins attached to electrical wires encased in an insulating material. The pins fit into an electrical socket (e.g. in a wall) and make an electrical connection.

A simple mould (usually made of wood) which is used in work in plastics is the **plug and ring mould**. The **plug** is the *male* part of the mould and the **ring** is the *female* part.

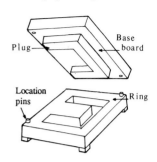

a plug and ring mould

Associations

See: **conductor, fastener, fixing, moulding, screw.**

Plywood

Pronounced: PLI-WOOD (*i as in fine, oo as in good*)

Origin

Ply is from the French *plier* meaning *to fold* or *to bend*. Plywood laminates were used in 2800 B.C. by the Ancient Egyptians, who bonded six plies together with wooden dowels. Plywood was first used for furniture by a German cabinet maker of Boppard-am-Rhein in 1819, who pioneered the steam-bent wooden chair with plywood seats in 1855. The production of aeroplanes during the 1914-1918 war, when plywood was used for many parts of the aeroplanes, advanced research into the use of plywoods. The first chair of moulded plywood, using synthetic adhesives, was made by two American architects, Charles Eames and Eero Sarrinen, in the 1940s.

using a panel saw to cut plywood

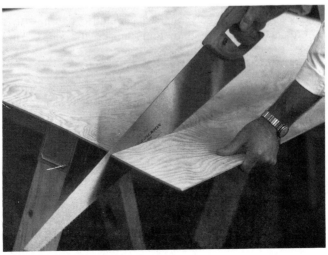

Meaning

Plywood is made by bonding three or more sheets of veneer (or *plys*) face to face, usually in a hot hydraulic press under high temperature and pressure, using synthetic resins, although cold presses are also used. The laminate is dried to have a moisture content of 8-15% and then sanded and trimmed to size.

Plywood is categorised by a number, which is determined by the the total number of sheets of veneer. The number can vary from 3 to 35. The direction of the grain of each sheet alternates, with the grain of each sheet being at right angles to the previous sheet. This arrangement produces a material which is strong in all directions. It is usual to have an odd number of layers, so that the grain direction is the same on the top and bottom surfaces of the piece of plywood.

Associations

See: **hydraulic, kerf, laminate, resin, synthetic, veneer.**

Polygon

Pronounced: POLY-GON (*o's as in got, y as in duty*)

Origin

From the Greek *polus* meaning *many* and *gonos* meaning *angle*.

Meaning

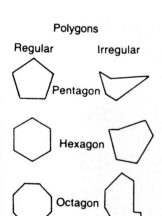

A polygon is a geometric figure with many (usually more than four) angles and sides. The following are polygons:

   tetragon.- a four sided figure which includes the square, rectangle, quadrangle, rhombus and parallelogram;

   pentagon - a five sided figure with five equal interior angles. (See the Pentagon Building in Washington, U.S.A.);

   hexagon - a six sided, six angled figure; (It is also called a **hexagram**. The star of David which appears on Israel's national flag is a hexagram.)

heptagon - a seven sided, seven angled figure;

octagon - an eight sided, eight angled figure;

nonagon - a nine sided, nine angled figure;

decagon - a ten sided, ten angled figure;

dodecagon - a twelve sided, twelve angled figure.

Associations
See: **angle, form, geometric, polyhedron, shape.**

Polyhedron

Pronounced: POLY-HEE-DRAN (*o as in got, y as in duty, ee as in see, a as in ago*)

Origin
From the Greek *polus* meaning *many* and *hedra* meaning *a base*.

Meaning
A polyhedron is a **solid** figure with many bases or faces. The following are polyhedrons:

tetrahedron - with four faces (bases) or triangles;

pentahedron - with five faces;

hexahedron - with six faces (e.g. a cube);

heptahedron - with seven faces;

octahedron - with eight faces;

decahedron - with ten faces;

dodecahedron - with twelve faces;

icosahedron - with twenty faces.

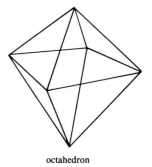
octahedron

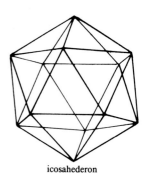
icosahederon

Associations
See: **angle, face, form, geometric, polygon, shape.**

Pored

Pronounced: PORD (*o as in port*)

Origin

From the Greek *poros* meaning a *passage* or *an opening through which things may pass*.

Meaning

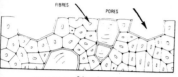
structure of hardwood

Pored woods are **hard woods** (botanically classified as *dicotyledons*), which have pores, or hollow, tubular cells (called also **vessels**), which run vertically in a tree to channel water from the roots to the leaves. The number of pores, their sizes and the pattern they make in growth rings differ from tree to tree and help to identify a timber. Their texture ranges from fine to coarse, depending upon the arrangements and size of the pores. Their grain is straight, sloping or interlocked.

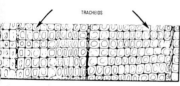
structure of softwood

Non-pored woods are **soft woods** (botanically classified as *conifers*). They do not have pores or vessels, and their cell structure is different from that of hard woods. Their texture is fine and their grain is usually straight, except in those species having numerous knots (e.g. radiata and cypress pines). The terms **hardwoods** and **soft woods** are misleading, as a number of hard woods are quite soft in composition (e.g. balsa) and some soft woods are relatively hard in composition and difficult to work.

Associations

See: **medullary, porosity, texture**.

Porosity

Pronounced: POR-OS-ITY (*1st o as in more, 2nd o as in loss, i as in pin, y as in duty*)

Origin

From the Greek *poros* meaning a *passage*. Eventually, it came to mean that an object which had passages in its material along which liquid could flow was *porous* and had *porosity*.

Meaning

Porosity means that an object has pores, like pores in human skin, which are minute holes (*passages*) through which fluids, air and gases may pass. Bricks, paper and sponges, for example, are porous. **Pinhole porosity** refers to a fault in metal casting where small holes appear in metal, owing to the escape of gas in the metal as it cools and shrinks.

Associations

See: **anodise, casting, filler, filter, finish, pored, priming, sealer, shrinkage.**

Positive

Pronounced: POS-A-TIV (*o as in loss, a as in ago, i as in give*)

Origin

From the Latin *positus* meaning *to place firmly* and then *positif* meaning *capable of being affirmed*; that is not negative. It came to mean to place something firmly in place, then to say something affirmatively. The use of *positive* in photography goes back to the invention of photography. Joseph Nicéphore Niépce (1765-1833), a Frenchman, was the first person to fix an image permanently when he made a photograph on a metal plate in 1826. His collaborator Louis-Jacques-Mandé Daguerre (1787-1851) produced the first negative, called a *daguerrotype*, in 1839. This was the first commercially-feasible process of obtaining photographic images on a metal plate. The method produced only one print at a time and it was obsolete by 1860, when the collodion (wet-plate) method was invented which allowed multiple prints to be made from one negative. In 1839, an English mathematician, William Henry Fox Talbot (1800-1877), perfected the callotype process of taking positive prints from negatives produced on glass.

Meaning

Positive refers to a photographic print or plate which shows light and dark values as they occur in the original image being photographed, black for black and white for white - the opposite to a **negative**.

It means also a model (carved, modelled or cast) from which a mould is made in order to reproduce the model form.

In electricity, positive refers to a point or electrode which relative to another point has a higher electrical potential.

Associations

See: **anodise, bromide, contrast, electrode, half-tone, moulding, negative, space.**

Preservation

Pronounced: PREZA-VASHUN (*e as in let, 1st a as in ago, 2nd a as in late, u as in fun*)

Origin

From the Latin *praeservare* meaning *to take care beforehand*, and *to save, to protect, to keep safe from harm and injury.*

Meaning

Preservation refers to the process of treating timber with a wood preservative to prevent its rotting and being attacked by insects (such as **pinhole borers** and **termites**) and consequently extending the timber's useful life. Such preservatives should be fire-resistant, colourless, odourless, poisonous to insects and fungi but not to humans and animals and should be cheap, durable and easy to apply. They should also have no adverse effect on metals (nails, screws, bolts etc.) or finishing substances, such as paint or varnish.

Nowadays, the most effective form of preservation treatment of timber is by a vacuum pressure-impregnation technique.

There are two kinds of wood preservatives in general use: *preservative oils* (such as creosote oil, "penta" (pentachlorphenol in mineral oil) and copper or zinc naphthenate) and *waterborne preservatives*, which are chemical salts dissolved in water (e.g copper, arsenic and chrome salts and boron and fluorine compounds). Timber is also preserved and protected by some pigmented finishes and also by some paints, varnishes and enamels.

moving timber into a preserving cylinder

Associations

To kyanize is to impregnate timber with a corrosive cleanser, such as mercuric chloride, to prevent decay and help preservation. The process was invented by J.H. Kyan in the early 1800's

See: **corrosion, develop, finish, paint, pigment, size, varnish.**

Priming

Pronounced: PRIM-ING (*i as in kite*)

Origin

From the Latin *primus* meaning *first*. Priming is the first operation in a task.

Meaning

Priming is the process of applying a *primer*, a first coat of finishing substance, which seals the pores of wood, and provides a **key** for the next coat of finisher. On external work, a primer usually contains a drying agent and a chemical to prevent corrosion.
Priming also describes the process of filling a pump intake with fluid to expel air and the injecting of petrol into an engine cylinder to facilitate starting.

Associations

See: **corrosion, finish, key, paint, porosity, size, varnish.**

Profile

Pronounced: PRO-FIL (*o as in go, i as in mile*)

Origin

From a now obsolete Italian word *profilare* meaning *to draw in outline* and *profilo* meaning *an outline*.

Meaning

A profile is the side view of a person's head and the outline of one side of a form or figure. It is also the form of a moulding in cross section.
A **profiling machine** is a milling machine where cutters can be directed to follow a pattern or profile.
profile paper is drawing paper covered with a grid, according to the scale used.

Associations

See: **contour, cross-section, grid, mechanical drawing, mill, moulding, pattern, shape, technical drawing, template**

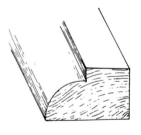

Profile of a wood bead moulding

Projection

Pronounced: PRO-JEK-SHUN (*o as in go, e as in let, u as in run*)

Origin

From the Latin *pro* meaning *forward* and *jectum* meaning *thrown*. Thus, to throw ideas forward or to plan or to place something so that it juts out and is conspicuous.

Orthographic is from the Greek *orthos* meaning *right* and *graphein* meaning *to write* and later *to draw*. *Orthogonal* is from the Greek *orthos* and *gonia* meaning *angle*. That is rectangular, with several right angles. *Isometric* is from the Greek *isos* meaning *equal* and *metron* meaning *measure*. That is *having equal measurements*. *Oblique* is from the Latin *obliquus* meaning *slanting, sidelong, indirect*.

Meaning

A projection, or pictorial view, is the way in which three-dimensional objects, having length, breadth and depth and six possible views (front, back, top, bottom and two sides), are represented on a flat surface, such as on paper.

An **orthographic or orthogonal projection** (sometimes called a *working drawing*) is a two-dimensional visual representation of a three-dimensional object, showing separately three views namely a **front view**, a **top view** and a **left side view**, which have correct scale and positioning, but where there is no convergence of parallel lines. It is probably the most useful projection in that it shows all the information related to shape, dimensions and arrangement of constituent parts.

Formerly these three views were referred to as *plan, front elevation* and *end elevation* and these terms may be found in some books.

An **isometric projection** is a representation on a flat surface of objects which are arranged in depth (with a 30° receding line) but where they are shown in equal scale with no reductions for how they might appear in perspective. Scales of height, width and depth are kept constant. It is useful in that by showing the length, width and thickness of an object it reveals proportion but it does not show the true shape of the object.

Oblique projections are produced from orthographic

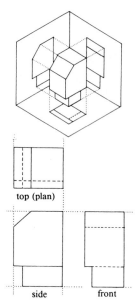

orthographic projection

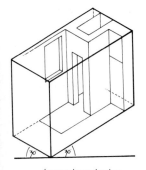

isometric projection

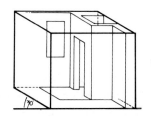
oblique projection

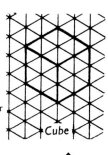
isometric drawing paper

front views with the addition and construction of parallel sides drawn at convenient angles of less than a right angle, usually 30° or 45°.

The most common method of orthographic projection used in industry is now the **Third Angle Projection**, which has superseded the **First Angle Projection**.

Associations

An **axonometric projection** is an orthographic projection, tilted to show three faces of an object from one viewing position.

Dimetric views (literally *two measurement views*) have two axes which are foreshortened an equal length and a third axis which is foreshortned a different length.

See: **angle, cross section, drawing, foreshortening, geometric, line, mechanical drawing, plane, perspective, foreshorten, technical drawing, Third Angle projection.**

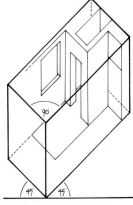
axonometric projection

Proportion

Pronounced: PRO-POR-SHUN (*1st o as in go, 2nd o as in por, u as in run*)

Origin

From the Old French *portionner* meaning *to share, give a part to*. Since the times if the Ancient Egyptians, designers have attempted to create what appears to the eye to be perfect proportion. The Ancient Greeks studied proportion, as did artists of the Renaissance, such as Leonardo da Vinci. The furniture of Robert Adams (1728 - 1792), the English architect and furniture designer, is noted for its dignity and superb proportion.

Meaning

Proportion refers to the relationship of the component parts of a building or of a constructed work to each other and to the total construction. Usually the word means that there is a correct relationship or ratio of

human proportions

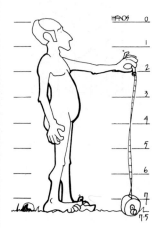

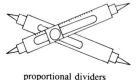

proportional dividers

one thing to another and that each component has been given its appropriate **portion**. It involves a comparison of elements in terms of size, height, width, shape, weight, quality placement etc.. Proportion may be expressed in comparative terms such as *brighter than, twice as large as*, etc.. Much attention has been given to proportion, particulary to the relationship of different parts of the human body. It has been found that normally the human body is about 7.5 times as high as the length of the head and that the total height is roughly equal to the width of the outstretched arms. The body is divided roughly into eight equal sections. It is generally considered that when proportion is produced a feeling of *harmony* and *balance* is achieved.

Associations

Proportional callipers have a pair of jaws at each end, one large, one small. A sliding screw allows adjustments, so that proportional settings can be made for reducing or enlarging work.

See: **callipers, distortion, elements, ergonomics, foreshortening, Golden Section, grid, scale, squaring up, symmetry.**

Pyrometry

Pronounced: PI-ROM-A-TRY (*i as in bite, o as in gone, a as in ago, y as in duty*)

Origin

From the Latin *pyra* from the Greek *pur* meaning *fire*, and the Greek *metron* meaning *measure*.
In 1730, a Dutchman, Petrus Van Muschenbroeck, invented a bar pyrometer. In 1750, a famous English ceramist, Josiah Wedgwood, invented silica cones for checking furnace temperatures and in 1782 he invented a pyrometer. Throughout the 19th century pyrometers were developed as thermo-electrical knowledge increased.

Meaning

A pyrometer is an instrument for measuring high temperatures (beyond the range of a mercury thermometer) in kilns and furnaces used for ceramics, enamelling, jewellery, glass and metal work. It uses a **thermocouple** which is a device for converting heat into electric cur-

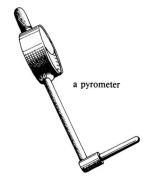

a pyrometer

rent, which can be measured to ascertain temperature. A **photo-electric pyrometer** is a device for measuring the temperature inside a kiln by using electrical currents produced by photo-electric cells. An **optical pyrometer** is a device which focuses on the heated area of a kiln from a safe distance and measures the intensity of light emitted from the kiln, which indicates on a gauge the temperature within the kiln.

The colour shown by the heat in the kiln is compared with the colour of a wire heated at a known temperature by an electric current. Kiln temperatures are also measured by **pyroscopes**, which are arrangements of ceramic minerals which melt at different specific temperatures. They are available in the form of cones (e.g. Seger cones), bars or rings. These can be linked to control switches, so that kiln temperatures can be controlled by switching the kiln heating system on and off. **Pyrometric cones** and other forms are made to bend, deform and melt at known temperatures. They do not measure temperature but how much heat-work has been done. Each cone is numbered, according to its melting point. The condition of the pyrometric objects can be seen through a peep-hole in the kiln and temperature adjustments can be made as needed.

Associations

See: **enamel, kiln, thermodynamics**.

Rake

Pronounced: RAK (*a as in make*)

Origin

From the Swedish *raka* meaning *to reach out* or the Old English *hraegan* meaning *to jut out*.

Meaning

The rake is the slope on the surface of a cutting tool (e.g. a drill's cutting face) or of the edges of the teeth of a saw. The rake angle must be negative when drilling plastic materials to ensure a scraping rather than a cutting action.

To rake is to incline or slope from the perpendicular to the horizontal.

Associations

See: **angle, bevel, cant, chamfer, drill**.

a raked wall.

Ratchet

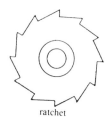
ratchet

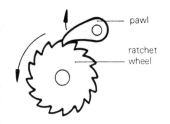
pawl
ratchet wheel

Pronounced: RACHAT (*1st a as in bat, 2nd a as in ago, ch as in church*)

Origin

From the French *rochet* meaning a *blunt lance-head, spindle or bobbin* and then a *rack* which is a metal bar with teeth on its edges which could work into the teeth on a wheel, pinion or screw.

Meaning

A ratchet is a device which allows a drive wheel or a bar, which has teeth on its edge (a **gear**), to turn or move in only one direction, as it is stopped from going in the opposite direction by engaging in one of the teeth of the wheel or by a bar, called a **pawl**. Ratchets are used on some braces, screw-drivers, fishing reels etc..

Associations

See: **jack, machine, pawl, pinion**.

Rebate

planing a rebate

using a rebate plane

Pronounced: REE-BAT (*ee as in see, a as in late*) OR RI-BAT (*i as in rib, a as in late*)

Origin

From Old French *raboter* meaning to plane which came from *abouter* meaning *to thrust against*.

Meaning

A rebate is a step-shaped channel cut at right angles into the edge or side or across the end of a piece of timber. It usually receives the edge or tongue of another piece of wood to form a **rebate joint**. Rebates, (or **rabbets** as they are sometimes called), are used for picture frames, window sashes, box construction, cheap drawers and for frameworks etc..
A **rebate plane** (or a **fillister plane**) and a **rebating saw** are used **to rebate** the edge of timber.

Associations

See: **groove, joint, plane**.

Reciprocate

Pronounced: REESIP-RA-KAT (*ee as in see, i as in lip, 1st a as in ago, 2nd a as in late*)

Origin

From the Latin *reciprocare* meaning *to move back and forward*.

Meaning

Reciprocating movement is motion backward and forward. A **reciprocating engine** is an engine in which pistons move forward and backward in a straight line. In engineering, **reciprocating motion** refers to the power which is transmitted from one part of a machine to another, as, for example, from a cam to its follower.
In photography, the **reciprocity law** states that *exposure = intensity X time*. *The exposure of a film is a product of lens aperture (intensity) and the shutter speed (time)*.

Associations

See: **cam, eccentric, machine, mandrel.**

Rectilinear

Pronounced: REK-TI-LIN-I-A (*e as in let, i's as in pin, a as in ago*)

Origin

From the Latin *rectus* meaning *straight* and *linea* meaning *a line*.

Meaning

Rectilinear means in, or forming, a straight line, or bounded by straight lines. Its opposite is **curvilinear**.

Associations

See: **drawing, elevation, form, geometric, mechanical drawing, shape, technical drawing.**

a rectilinear design

Relief

Pronounced: RI-LEEF (*i as in wrist, ee as in see*)

Origin

From the Italian *relievo* meaning *raised* and the French *relever* meaning *to lift* or *raise* (compare with a *lever* which helps to raise something).

Meaning

Relief refers to a design which is carved or cut out by some means (e.g. by a tool or by a chemical), so that the design, or part of it, is raised and stands out from the background and surface on which it is set. **Bas-relief** refers to work on a flat or curved surface, which is in **low-relief**. That is, there is little projection from the surface, as on a coin. This is in contrast to **mezzo-relief** where the projection from the surface shows half the full form of the figure or object, or **alto-relief**, which is **high-relief**, where the design contacts the surface at only a few points.

It is also a term in machining to describe the clearance on cutting tools to stop metal from rubbing behind the tool's cutting-edge

A *bas-relief* effect can be achieved on photographs by combining negative and positive films of the same image where they are placed slightly out of register.

Associations

Torentic refers to carving, chasing or embossing in relief. **Anaglyph** refers to a very low-relief sculpture or embossing.

See: **chasing, clearance, composition, design, emboss, engraving, intaglio, machine, repoussé.**

relief work on a chair leg

Replica

Pronounced: REP-LI-KA (*e as in met, i as in pin, a as in ago*)

Origin

From the Italian *replicare* meaning *to reply* or *say again*. It then came to mean *do again*.

a fax machine

photograph by courtesy of NEC Pty Ltd

Meaning

A replica was originally an exact copy of a work . Nowadays, it can mean one of a number of versions (duplicates) of a work or of a reproduction of any document or workpiece.

A **facsimile** is a reproduction of a drawing, print, artefact or manuscript. Originally, it meant a precise copy in all respects, including the materials used. Nowadays, it usually refers to copies of original documents or graphics by the use of an electronic reproduction machine, one of which is called a *facsimile machine*, which is often abbreviated to a **fax**.

Associations

See: **cire perdu, facsimile, motif, pantograph, pattern, squaring up, symmetry**.

Repoussé

Pronounced: RE-POOZ-A (*e as in return, oo as in moon, a as in late*)

Origin

From the French *pousser* meaning *to push* or *pound* and the prefix *re* meaning *from behind*. *Repousser* means *to push back* or *to push from the back*. Before casting was invented (probably in Egypt about 2300 B.C.) all metal work was produced with a hammer. Decorative metal work, including repoussé work, especially in precious metals, continued to be made with a hammer.

Meaning

Repoussé refers to decoration in relief on metal (i.e. jewellery, silversmithing and goldsmithing), leather or enamelling. It is an alternative term for **embossing**. The metal used must be able to stretch when hammered and should be less than 0.4 mm. thick before it is worked on. Excellent metals to work on are sheet pewter, copper and gold. It is produced by hammering (sometimes called "beating up") with punches (of which there is a variety, made of shaped brass or steel) or pressing mainly on the reverse side of the object to raise designs from the surface of the object or material in relief. In metal, the work is mounted on a bowl of pitch, lead or wood to provide the necessary resilience

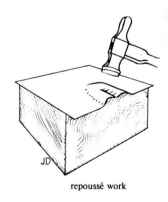

repoussé work

a repoussé hammer

and elasticity for the stretching of the metal. The repoussé work is usually finished off by **chasing**, which defines the repoussé work.

Associations

Drawing is the term used in metal work for the stretching and shaping of metal by hammering and heating it.

See: **burr, chasing, drawing, emboss, hammer, impression, pitch, relief.**

Resin

Pronounced: REZIN (*e as in let, i as in fin*)

Origin

From the Latin *resina* which described an adhesive substance which exuded from some plants and some trees, such as the fir and the pine.

Meaning

In plastics, resin is any thick natural or synthetic (often called **resinoid**) liquid which will polymerise and cure as a thermoset or a thermoplastic. The resins usually used in workshops are polyester resin (used often in fibre glassing) and epoxy resin (used with a hardener mostly as an effective adhesive for a wide range of materials and also in fibre-glass work).

Resins, which can be natural or synthetic, are the major ingredient in varnishes, lacquers and plastic finishes which produce a smooth surface film.

Resin hardness is measured on a scale of 1 to 6, where 1 is the hardest.

Associations

Acetone can be used for removing uncured polyester resin from brushes, clothes, equipment . Resins should not be sanded on a sanding disc as the particles are a health hazard.

See: **accelerator, adhesive, catalyst, conductor, filler, finish, harden, knot, lacquer, paint, particleboard, plastic, plywood, sculpture, synthetic, thermoplastic, thermosetting, varnish, viscosity.**

Resist

Pronounced: RI-ZIST (*i's as in wrist*)

Origin

From the Latin *resistere* meaning *to stop*.

Meaning

Resist refers to a material or chemical (e.g. varnish, rubber cement, masking tape, opaque pigment, floor paste, wax, hard ground) which is used to mask or block out part of a work so that the covered part does not attract a dye, ink, glaze, paint or chemical that is being used in the process. This is called a **resist technique**.

Associations

Resist is also referred to as **stopping out**.
See: **engraving, etching, mordant.**

Rip

Pronounced: RIP (*i as in dip*)

Origin

Probably from the Frisian *rippe* or the Flemish *rippen* or the old German *reppen* meaning *to tear off, to rip, to strip off roughly*.

using a rip saw

Meaning

To rip is to saw along the length of the grain of thick timber. A **rip saw** is a large saw with teeth like small chisel heads. Most large pieces of timber are ripped by a power-driven circular-saw (portable or fixed) with a **ripping fence** as a guide. A **push stick** is used to push narrow or smaller lengths of timber into the teeth of the circular saw from a distance, as a safety precaution.

Associations

Cutting across a piece of timber is called **crosscutting**. A **ripper** is a tool made from a scrap of sheetmetal which can be used to scribe lines parallel to an edge. See: **grain, scribing, saw.**

Rivet

types of rivets

Pronounced: RIVAT (*i as in give, a as in ago*)

Origin

From Old French *river* meaning *to clench* which is *to grasp firmly* or *to secure*.

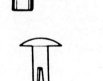

cup, round or snap rivet

countersunk rivet

Meaning

A rivet is a malleable metal pin consisting of a shank and a shaped head. The diameter of the shank should be equal to twice the thickness of the metal being riveted. It is usually made from mild steel, copper or aluminium. Riveting is a method of joining metals or other materials together using rivets. A rivet passes through a hole drilled into two overlapping materials and each end of the rivet is flattened using an engineer's ball-pene hammer. This process (which is called **clinching**) bonds the materials. A **bifurcated rivet** (*two-pronged*)) is a rivet with split shanks which is used for joining fairly soft materials, such as rubber or leather. The shanks are turned up and hammered so that they spread out under the material being rivetted. A **blind rivet** is one that can be fitted from one side only. Using special rivetting pliers or a rivet gun, a pin is either pulled with the pliers or pushed with a gun through the hollow shank of the rivet which expands the shank so that it grips the surrounding material. **Roves** are copper washers used with rivets to hold one metal to a softer metal or to join soft metals together.

bifurcated rivet

rivet and rove

210

Associations

See: **bolt, fastener, fixing, hammer, joint, lap, malleable, pene**

Rout

Pronounced: ROWT (*ow as in cow*)

Origin

To rout is a variation of *to root* which is from the Old English *rot* meaning *a swine's snout*. To rout came to mean to grub something out with the snout and later to furrow or to make a groove.

Meaning

To rout is to make grooves or trenches of various shapes in wood of a uniform depth.

A **Router** is a cutting tool with a blade which has two handles attached. It is used for smoothing the surface of grooved parts of a wood workpiece. A portable **router** is a device with a high-speed electrical motor which is mounted on top of a circular base, which has a chuck which can hold bits of various kinds. Routers can use either a spade-ended blade, which is used in sharp corners where it is difficult to use a chisel or a chisel-ended blade, which is a general-purpose blade. The router is very versatile and can be used with appropriately-shaped bits for grooving, rebating, trenching, lapping, mortising and tenoning, edge-moulding, chamfering, inlay recessing and low-relief carving. A **router plane** is a special plane used for cutting and smoothing the bottoms of rectangular grooves.

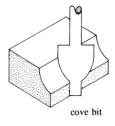
cove bit

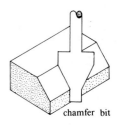
chamfer bit

two of the various router bits

Associations

See: **bit, chamfer, chuck, groove, inlay, moulding, plane, relief.**

using a router planer

Saw

Pronounced: SAW (*aw as in draw*)

Origin

From the Old English *sagu* meaning *a cutter*.

Hand saws were used in Ancient Egyypt, Greece and Rome before the 7th century B.C. The first raked-teeth saws were introduced in about 1490. The first circular saw was invented in 1777 by S. Miller of England. and numerous other circular saws were invented in the early 1800s. The first effective band saw was developed by a French woman, Madame Crépin in 1846. This was followed by a number of improved versions during the first half of the 19th century.

circular saw

Meaning

A saw is a cutting tool formed of a blade, band or disc of thin, high-quality steel with a toothed (*saw-toothed*) edge. Some modern saws have teflon-coated blades, which reduces friction. The majority of saws have adjacent teeth bent out (a third of the length of each tooth) from the saw blade in opposite directions. This is known as a **set**. The set provides a clearance for the blade and helps to prevent its jamming when cutting, as does the tapering and skewing back of blades on good-quality hand saws.

Saws are classified according to a *point* system. The teeth are measured by the number of points per 25 mm. A ten point saw, for example, has nine teeth to the 25 mm. A 14 point saw has smaller and more teeth than a 6 point saw and produces a finer cut. Nowadays, there are many saws available, each of which has a particular function in the cutting of wood, metal or plastic, ranging from fine-toothed **hacksaws** for cutting metal to **hand saws** (e.g. a **rip saw** and a **panel saw**), **back saws** (e.g. a **tenon saw** and a **dovetail saw**) and **curve-cutting saws** (e.g. a **coping saw**, **bow saw**, **keyhole** or **pad saw** and **compass saw**) for cutting wood.

A **band saw** is a power-driven saw which has a flexible, endless steel blade running as a belt between two wheels. It cuts both straight and curved shapes. A **hole saw** is a circular saw available in a range of diameters, which is used with a drill to cut holes in metal, wood and fibre.

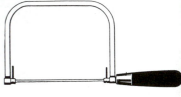

Coping saw.

Associations

A **saw file** is a three-cornered file used for sharpening saws. A **saw set** is an instrument for turning the teeth of a saw alternately right and left. That is *setting the saw*.

See: **clearance, cross section, fretted, hardboard, jig, kerf, mitre, panel, pitch, rip, taper, veneer.**

using a saw set

Scale

Pronounced: SKAL (*a as in late*)

Origin

From the Latin *scalare* meaning *to climb*. Scale shows how something *climbs* in size. Note the English word *to scale*, which means *to climb*, as in *to scale a wall*.

Meaning

Scale indicates a measurement and shows the relative proportion of one thing to another. Something drawn on a one quarter scale is one quarter of the size of the original. Things designed can readily be reduced or increased in scale. Most working drawings prepared by a designer or draughtsman are drawn to scale, so that

a scale

213

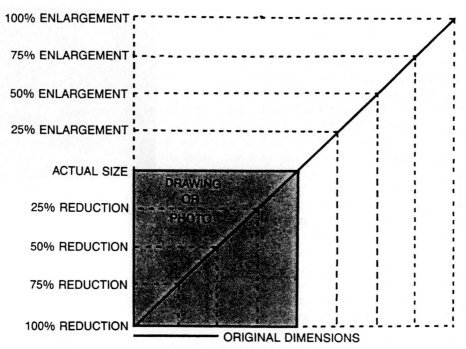

enlarging or reducing

large jobs can be be proportionally reduced and presented on paper. These are called **scale drawings**, where the scale is given as a ratio, such as 1:100, as used in house plans, or 1:250,000 as used on some maps.

Scale also describes a thin, black layer of oxide which appears on metals when they have been heated. It should be removed before the metal is painted or subjected to a finishing process.

An **I.S.O.(International Standards Organisation) scale** is used for measuring how quickly a photographic film is affected by light. The higher the number, the less light is required to expose the film correctly. Another similar photography scale is the D.I.N. (Deutsche Industrie Norm - meaning German Standards Organisation). A **grey scale** in photography is a chart of grey tones starting with white and progressing in grades of tone to black. The chart is used as a test or guide in photographic processes. **Mohs' Scale** is a ten point scale of hardness for metals (devised by Friedrich Mohs, a 19th century mineralogist). It measures the ability of one metal to scratch or be scratched by another. An unknown substance is compared with a metal whose hardness is well known. Silicon carbide will scratch tungsten, so silicon carbide is the harder metal.

Associations

A **triangular scale** is a triangular shaped rule which combines several scales. A **diagonal scale** is a means whereby a length can be accurately divided into a large number of small parts.

See: **callipers, dip, drawing, exposure, gradation, grid, hardening, mechanical drawing, geometric, grid, indexing, lens, pantograph, perspective, profile, proportion, squaring up, symmetry, thicknesser, torque.**

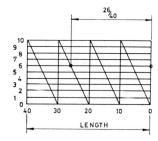

diagonal scaling method

Scantling

Pronounced: SKANT-LING (*a as in ant, i as in sing*)

Origin

From Old French *escantillon* meaning *a small piece, a sample* or *a small piece of wood or stone.*

Meaning

A scantling is a timber beam of small cross-section (about 5 to 10 cms thick and 5 to 12 cms wide) or the size to which timber or stone is to be cut.

Associations

See: **flitch.**

Scarfing

Pronounced: SKARF-ING (*a as in far*)

Origin

From the Norwegian *skarv* meaning *a joint.*

Meaning

To scarf is to join two pieces of timber end-wise, so that they make one continuous piece of timber. A **scarf joint** is made by overlapping two pieces of timber or metal that will fit each other and then bevelling or notching them before bolting, brazing or screwing them together. It is also the v-shaped cut made in a tree with an axe when a tree is to be felled.

In forging, scarfing is the tapering or bevelling of the ends of two pieces to be joined so that a smooth rather than a protruding joint is produced.

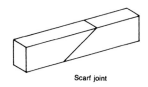

Scarf joint

Associations

See: **bevel, bolt, brazing, lap, joint, notching, screw, taper.**

Scoring

Pronounced: SKOR-ING (*o as in for*)

Origin

From the Old Norse *skera* or *skur* meaning *to cut* or *shear*.

Meaning

To score is to cut a line or groove with a sharp tool (e.g. an **awl**) or by burning into glass, plastic, jewellery, ceramics, leather or wood, or to make a furrow in paper (usually using a blunt instrument), in order to help in folding the paper without it tearing.

Associations

A **scorer** is a tool with a rounded end, used in engraving. It used for making thick lines and for removing areas of an engraving plate or wood block.

See: **chasing, engrave, groove, line, wood block.**

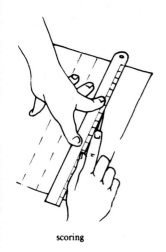

scoring

Screw

Pronounced: SKREW (*ew as in new*)

Origin

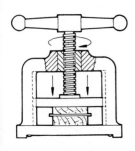

The origin of screw is obscure. It may have come from the Latin *scobem* meaning a *hole* or from the Old French *escroue* meaning a *nut or female screw*.

A Greek mathematician, Apollonius of Perga, worked out the geometry of the spiral helix around 200 B.C.. Archimedes (287 - 212 B.C.), also a Greek mathematician, is reputed to have invented the screw and applied it to practical activities, such as a screw to raise water and another to drag a loaded ship onto dry land. Wooden screws were used in Ancient Rome and they continued to be used in many countries until metal screws became relatively cheap in the 19th century. The first-known description of a metal screw was in a treatise written by a German mining engineer, Georgius Agricola, in 1556. Jacques Besson, a Frenchman, is credited with the invention of the first practical screw-cutting machine in 1586. Hand-made metal screws were very expensive, consequently they were used only to join wood to other materials (e.g. in the making of guns). Joints and dowels were used for joining wood to wood. The first screw-making machine, which produced blunt-ended screws, was patented by Job and William Wyatt in England in 1760. In the 1840s, the British engineer, George Nettleford, produced a machine which produced pointed screws. In 1860, James Whitworth's standardisation system for screws (which he invented in 1841) was adopted in England. His thread (the B.S.W. - the British Standard Whitworth) had a thread angle of 55° and each diameter of screw had a number of threads per inch. In the late 1800s the British Standard Fine thread (the B.S.F.) was adopted, where the number of threads per inch was increased. In the U.S.A., M. Sellers invented the American Standard thread, which had a thread angle of 60°. It was officially adopted in 1868. Nowadays, the ISO (International Standards Organisation) metric thread is universally used. It has a thread angle of 60° and two pitches of threads: *coarse* and *fine*. The fine series of threads is used where great tightness is essential and where vibration is a factor.

An Englishman, C.W.Parker invented an automatic screw machine in 1879.

Meaning

A screw is a metal cylinder with a spiral groove (the **thread**) on either its outer surface (called a *male screw*) or its inner surface (called a *female screw*). It is used as a fastener (e.g. for wood or metal) or as a mechanical power (e.g. a screw propeller or screw-press). A screw thread may be left-handed, where the screw rotation is anti-clockwise, or right-handed where the screw rotation is clockwise. The distance a screw advances after one complete turn is called a **lead**. Screws are usually smaller than bolts and are more varied in shape and function.

A screw (and a bolt) has three main parts: *the major diameter* which is the unthreaded part of the screw, the *minor or root diameter*, which is the diameter of the screw at the bottom of the thread and the *pitch* which is the distance from thread to thread.

Wood and metal screws are usually named according

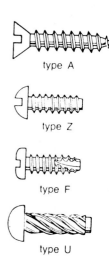

type A

type Z

type F

type U

self-tapping screws

a coach screw

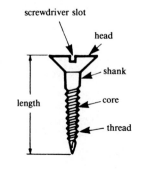

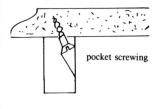
pocket screwing

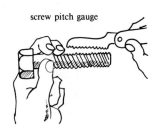
screw pitch gauge

to their shape such as: round head, square head, cheese head, raised head, hexagonal head, square (or coach) screw, countersunk head and phillip's head. Wood screws are usually made of mild steel, copper or brass (Mild steel soon corrodes in European oak.), and some are **galvanised** (coated with zinc) to prevent corrosion, or coloured by plating or oxidisation to match the colours of the fittings they secure

A **screw-bolt** is a bolt threaded at one end to take a nut, which has a female screw. A **screw-driver**, which fits into a slot in the head of a screw (or the two slots of the four-point star-shaped recess in the head of a phillip's screw), is used for driving screws into, or loosening them from, materials. A **stud** is a member threaded at both ends.

Associations

Slot screwing is a method of inserting screws so that the screwheads are not seen. **Pocketing** is a method fitting screws when joining **rails** to a table. A **tap** is a tool used to make an internal thread. A **thread pitch gauge** is a gauge for measuring the number of threads per 25 mm on a screw. **Backlash** describes the looseness between a threaded shaft and a nut owing to excessive wear.

See: **bolt, counterbore, countersink, die, fixing, gauge, injection moulding, jack, jig, lathe, oxidation, pitch, plug, scarfing, threads.**

Scribing

Pronounced: SKRIB-ING (*i as in pine*)

Origin

double edge engineer's scriber

From the Latin *scribere* meaning *to write*.

Meaning

Scribing is marking out accurate lines on a hard surface, e.g on hardened plastic or metal. The **ring scriber** and the **engineer's scriber** are instruments most often used to make permanent lines on smooth surfaces.

Associations

An **awl** is a sharp, pointed tool with a wooden handle. It is used for scribing and marking holes.
See: **line, rip.**

Sculpture

Pronounced: SKULP-CHA (*u as in bull, cha as ture in picture*)

Origin

From the Latin *sculptura* from *sculpere* meaning *to carve*.

Meaning

Sculpture is the art of producing forms in three dimensions through the elements of volume, space, movement and materials. Until the 20th century, there were two types of sculpture. One is the **subtractive** technique where material (e.g. stone) is carved and the material is reduced until the desired form is produced. The other is the **additive** technique where a pliable material (e.g. wax or clay) is built up until a desired form is produced. In the 20th century, artists (e.g. Picasso) have used several mediums to construct forms. The single most important technique in modern sculpture is that of **welding**. Today sculptors use not only stone, clay, wood, metals and glass-fibre but also plastics, resins, in fact any materials, including junk.

wood sculptures

Associations

See the works of Umberto Boccioni (1882-1916), Constantin Brancusi (1876-1957), Jean Arp (1886-1966), Auguste Rodin (1840-1917), Naum Gabo (1890-1977), Alberto Giacometti (1901-1966), Barbara Hepworth (1903-1975), Henry Moore (1898-1986), Jacob Epstein (1880-1959), Marino Marini (1901-1980), Julio Gonzales (1876-1942).

See: **casting, chisel, cire perdu, form, laser, plastic, resin, shape, tooling, welding**.

Sealer

Pronounced: SEELA (*ee as in bee, a as in ago*)

Origin

From the Old English *seel* meaning *an image, figure or mark*. A seal in wax (often from a ring) was put on important letters to fasten them and ensure they were opened only by the person to whom they were addressed. The word seal came to mean *to fasten firmly* or *enclose*.

Sealer and *sealant* are modern derivations.

Meaning

A sealer is a substance which is applied to wood or other porous materials ,in order to close the pores in the substance. If this were not done, an excessive amount of varnish, paint or other finishing materials would soak into the porous material, making the finishing process much more expensive than it need be. Sealers used under lacquers and plastic finishers usually have a plastic base.

Associations

See: **finish, plastic, porosity, size**

Seam

Pronounced: SEEM (*ee as in see*)

Origin

From Old English *seam* from *siwian* meaning *to sew*. A seam came to mean *that which is sewn*.

Meaning

A seam is the line formed by the sewing together of two pieces of material.

In metalwork, seaming is the joining of two pieces of sheet metal by folding the edge of one piece over the edge of the other. A **sealing machine** is a power tool for bending sheetmetal. A **seaming iron** is used for setting sheetmetal work seams.

Associations

See: **joint, key, lap.**

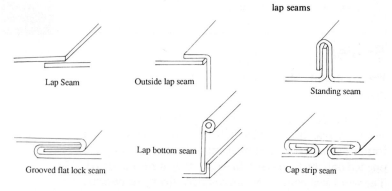

Seasoning

Pronounced: SEEZ-ANING (*ee as in see, a as in ago*)

Origin

From the Old French *seson*, which came from the Latin *satio* meaning *seedtime*. One of its meanings came to be *to take action at an appropriate time to ensure a product is preserved or fit for use.*

Meaning

Seasoning is the process of drying out **green wood** (timber which still has its sap), so that it is ready for use. The seasoning makes the timber lighter, harder, more durable, dimensionally more stable (although a little change across the grain may occur even after correct seasoning), easier to saw, plane, chisel and drill and also better able to take surface finishes. The timber is seasoned when the amount of moisture or sap in the timber is about the same as the moisture in the environment (its humidity) in which the timber will be used. This is called the **equilibrium moisture content** stage. As timber has about half its weight in moisture when it is cut, there must be a controlled drying-out process. This is done in one of three ways: by drying the wood naturally in the open, which may take six months; by drying it with hot air and steam in a timber kiln, which may be completed in three or four days; or by a combination of the open-air and kiln methods. The open air method is sometimes preceded by what is called **water seasoning**, where the timber is placed in running water for about three weeks. During this time much of the timber's sap is washed out.

air seasoning of a stack of timber

Associations

See: **distortion, hardening, hygroscopic, kiln, shrinkage, warp**.

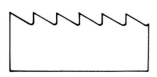
serrated teeth of a rip saw.

Serrated shaft

Serration

Pronounced: SE-RASHAN (*e as in set, 1st a as in late, 2nd a as in ago*)

Origin

From the Latin *serratus* meaning *saw-like*. The Latin *serra* meant a *saw*.

Meaning

Serrations are notches or grooves, shaped like teeth. A serrated engraving tool can produce a hatching effect.

Associations

See: **engrave, groove, notching**.

Shape

Pronounced: SHAP (*a as in late*)

Origin

From the Old English *sceapan* meaning *to form* or *to make*.

The Ancient Greeks and Romans used elaborate shapes for their mouldings. The curves of the Greek mouldings were based on conic sections, namely the ellipse, parabola and hyperbola. The *volute*, a spiral scroll, (from the Latin *volvere* meaning *to roll*) was the chief design feature in Ionic and Corinthian capitals of Ancient Greek architecture.

Ancient Roman mouldings were based on the arcs of circles. Shapes used during these times were: cyma, cavetto, ogee, ovolo, scotia (a concave curve, also called a cavetto) and torus (a convex curve, the reverse of scotia).

Modern design has rejected elaborate shapes for decoration, preferring chamfers, bevels and round edges, which are much easier to keep clean.

The first **shaping machine** is thought to be that invented by an Englishman, James Nasmyth, in 1836.

Meaning

Shape refers to the appearance of an object rather than to its structure. The term is often used for **form**, although to be correct, shape is only one of the elements of form (which includes proportion, scale, mass, volume, structure and texture). Shapes can be three-dimensional (e.g.

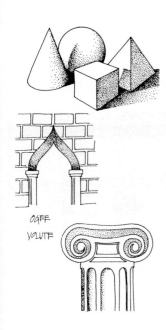

222

a person's egg-shaped head) or two-dimensional (e.g. a photograph of a person's oval head). Shapes are associated with human emotions or ideas. For example:
- a circle or tondo suggests something completed or perfect or at rest;
- jagged shapes suggest pain or tension;
- horizontal shapes suggest calm, peacefulness and rest;
- vertical shapes suggest uprightness and nobility;
- diagonal shapes suggest movement and dramatic action;
- pyramid or triangle shapes suggest stability (having a large base) and permanence.

There are hundreds of two and three dimensional shapes and many words to describe shapes, for example: furciferous-fork shaped; hamiform-hooked shaped; helicoid-screw shaped; obconic-pear shaped; palmate-shaped like a human hand; scutiform-shaped like a shield; uniform-J shaped; zygal-H shaped; sigmoidal-curved in two directions; stelliform-shaped like a star; campanulate-bell shaped; cordate-heart shaped.

Shape in engineering is a term used for rolled (made by a rolling process) structural metal, such as beams, angles and a variety of steel bars.

A **shaper** or **shaping machine** is a power-driven machine tool which is used for the planing of flat surfaces, especially where mass-produced work is required. It can cut grooves, slots etc. in solid metal. It comprises a reciprocating ram which holds a tool head, which, in turn, holds a cutting tool which is adjustable to cut horizontally, vertically or at any angle.

Associations

A **swage** is a tool used to shape metal. Work held on a swage is struck and shaped with a hammer. **Stamping** is a process of shaping sheet metal by forcing the metal into dies which create impressions. **Metal spinning** is a method of cold forming flat discs of metal into three-dimensional shapes on a lathe. A variety of spinning tools can be used, such as a round-nose forming tool, a ball tool, a trimming tool, a tongue tool and a beading tool. Sheet metal can be shaped by **hollowing** (to produce a hollow or spherical shape by hammering the sheet metal from the inside), by **sinking** (by hammering the metal from the inside and stretching it to form a flat-bottomed object), by **raising** (by hammering the metal from the outside while it is on a stake) and by

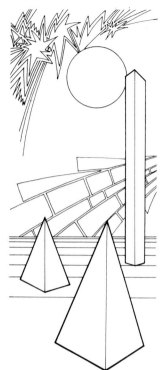

shaping wood with a file

planishing (see: **plane**).

A **spokeshave** is a tool with a cutting blade which is used for shaping and smoothing the edges of wood. **Cabriole** is a design in furniture where table and chair legs are plump and convex at the kneee and are slim and concave at the ankle.

See: **abrasive, blow moulding, burr, chisel, contour, drawing, die, elements, figure, forge, form, hammer, lathe, lens, line, malleable, mechanical drawing, mill, moulding, paring, pene, polygon, polyhedron, profile, sculpture, swage, symmetry triangulation, vulcanise, whetting.**

Shear

Pronounced: SHEAR (*ea as in fear*)

Origin

From Old English *sceran* meaning *to cut* or *to separate*.

Meaning

To shear is to cut sheetmetal by two blades passing each other (e.g. as in a bench shear, a guillotine, tinsnips or Tinman's curved snips, aviation shears, universal snips or jeweller's snips). A **bench shear** or **lever shear** is a machine for cutting sheet metal, where a sharp, knife-like blade drops perpendicularly onto metal held in place and which then cuts the metal along prescribed lines. The action of the cutting blade can be operated by hand, foot or by a power source. A **cropper** is a shearing machine used for cutting off (*cropping*) such things as rods.

Snips is another term used for hand shears used in sheet-metal work.

Associations

See: **crop**.

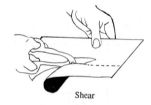
Shear

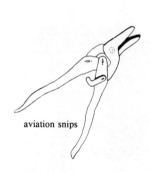
aviation snips

Shrinkage

Pronounced: SHRINK-IJ (*i as in ink, ij as idge in bridge*)

Origin

From the Old English *scrincan* meaning *to wrinkle* or *shrivel*.

Meaning

Shrinkage is contraction of a material. Timber shrinks owing to a loss of moisture; metal shrinks (or contracts) with a loss of heat. The percentage of shrinkage in timber (the dimensional reduction from the green timber state to the dried state) varies according to the species of wood and the method of drying. Timber tends to shrink in width and thickness rather than in length. Shrinkage can be controlled to some extent by treating the timber with chemicals (such as salt, urea, ammonium sulphate or polythylene glycol). Timber continues to shrink until the moisture or sap content in the timber is about the same as the moisture content (the humidity) of the air where the timber is. Even then there can be minor shrinkage and swelling changes in the timber, depending on the environmental temperature and humidity. Irregular and excessive shrinkage in different parts of timber while it is being dried leads to distortion (bowing, cupping, twisting and end splits) and even to **collapse**.

Allowance is made for some shrinkage in the making of furniture. Shrinkage buttons, slot screwing and panel grooves which are deep enough to allow movement are some devices used when making such things as door panels, table tops, drawing boards.etc.

Shrinking-on is a process in metalwork of fastening two cylindrical pieces (e.g. pipes) together. An outer piece is heated so that it expands. While hot, it is passed over a cold inner piece and as it contracts or shrinks, it bonds onto the inner piece to produce a **shrink fit**.

SHRINKAGE IN TIMBER

Associations

See: **distortion, fit, hardening, hygroscopic, kiln, porosity, seasoning, warp.**

Size

Pronounced: SIZ (*i as in rise*)

Origin

From the Middle English *sise* meaning *setting or fixing with a gluey substance*.

Meaning

Size is a gelatinous (i.e. it turns to gelatine or jelly when cold) substance,such as resin, glue or starch. Originally, it was made from the bones or hides of animals. It is put onto porous materials to seal the pores in them. This prevents the material from absorbing paint or other finishing substance. Some fabrics are sized (usually with a chalk-starch filler) to give them strength, called **body**. Size, in the form of paraffin wax, alum, asphalt or resin, is used in the making of particleboard or fibreboard to increase their water resistance.

Associations

See: **filler, finish, mordant, particleboard, preservation, porosity, priming, sealer**

Sketch

Pronounced: SKETCH (*e as in let, tch as in stretch*)

Origin

From the Greek *schedios* meaning *sudden* and then the Latin *scheduus* meaning *made in an off-hand way*.

Meaning

A sketch is a rapid but neat draft or outline of a composition which is made quite spontaneously. It is a designer's first effort to put down on paper his or her first impressions. It is a trial run. It is different from a **study**, which provides more detail.

Sketching (or drafting) is an essential stage in any design work, as the designer or engineer considers in sketch form a variety of ways of dealing with a design problem before a viable, economically-feasible and aesthetically-acceptable solution is reached.

The term **roughing out** is sometimes used for sketching.

Associations

See: **design, draft, drawing, elevation, grid, mechanical drawing, technical drawing.**

Solder

Pronounced: SOLDA (*o as in old, a as in ago*)

Origin

From the Latin *solidare* meaning *to fasten* or *solidus* meaning *solid*. That is *to make solid and firm*.

Metal workers in the Middle East discovered the method

of making lead-tin solder (**hard solder**) at about the same time as they discovered **bronze**, about 3500 B.C.. **Hard solder** of gold and silver was known about 2500 B.C.

Meaning

A solder is a mixture of metals (termed an **alloy**), which has the ability when it is in a molten state to stick strongly to other metals. When it cools it sets hard and is able to fuse and bond the edges of metals together. There are various kinds of solder for the different bonding of metals but the most common kind (**plumber's solder**) consists of equal parts of lead and tin. This melts at a relatively low temperature (about 250°c) and is called a **soft solder**. Solder has a melting point lower than the metals to be joined. **Flux** is always applied to the solder and to the metal to help the solder to flow more readily.

To solder means to join two pieces of metal (which do not themselves melt) together using solder which is heated until it is molten, using a **soldering iron**, a **soldering bit** or a **soldering gun**.

a soldering gun

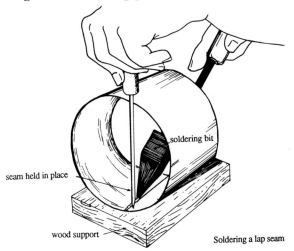

seam held in place

soldering bit

wood support

Soldering a lap seam

Soldering is much used not only in metalwork but also in stained-glass work. In jewellery work, silver solders or gold solders are used which melt at high temperatures (about 700 to 800°c), using a gas torch. This is called **hard solder**. **Sweating** is a soldering technique where two metals are coated with solder, clamped together and heated, so that the solder melts and the metals are fused.

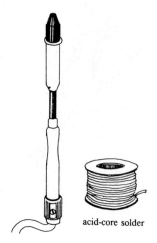

a soldering iron

acid-core solder

Acid-core solder is a wire solder which has a core of flux. As the solder melts, the flux removes any oxide film from the metals being joined and so helps to bond the metals.

Retinning is the process where a soldering bit is reconditioned. The bit is heated to loosen the old solder, filed, and cleaned with a pickle solution if necessary, and then coated with a new layer of solder.

Associations

Brazing is the joining of two metals, using brass or a similar copper-zinc alloy and a flux to make the joint. The metals are bonded at a temperature above 430°c. As the metal parts are heated, the brazing alloy is drawn into the joint by capillary action. Brazing is often used to join dissimilar metals, such as brass to wrought iron or copper to steel.

See: **alloy, bit, dip, enamel, filigree, fillet, flux, joint, pickle, oxidation, weld.**

Solvent

Pronounced: SOLV-ENT (*o as in got, e as in token*)

Origin

From the Latin *solvere* meaning *to loosen* or *dissolve*.

Meaning

A solvent is a liquid which can dissolve certain materials, such as water for acids and alkalis, terpentine for oil paint, wax and varnish, acetone for polyester resins, methyl acetate for lacquers and methyl acetone for rubber. Solvents are often used in solutions of different substances where the solvent will evaporate, leaving just the substance one needs. For example, **benzol** is part of some rubber cements and it evaporates as soon as the cement is applied A solvent which evaporates very quickly is said to be **volatile** (e.g. petrol, methylated spirits and chloroform). Solvents are used extensively for cleaning purposes.

Associations

Thinners is a solvent used to thin varnishes, lacquers and paint. A solvent applied to a plastic material will dissolve the plastic and make it tacky. Another piece of plastic will adhere to the tacky surface.

See: **lacquer, paint, plastic, varnish, vehicle**

Space

Pronounced: SPAS (*a as in late*)

Origin

From the Latin *spatium* meaning *extent, room, open area*.

Meaning

Space is a term which is used in design to indicate area or shape. Two-dimensional space (a plane) has height and width; three-dimensional space (volume) has height, width and depth. Positive (or occupied) space indicates a filled area or shape; negative (or unoccupied) space indicates an open, blank area or shape, surrounding a positive shape. The illusion of space (and depth and distance) can be created by: overlapping of objects, linear perspective, aerial perspective, differing size relationships, transparency, interpenetration, the contrast of sharp and diminishing detail, advancing and receding of colours, and changes in colour and texture values.

Associations

See: **colour, elements, geometric, negative, perspective, plane, positive, shape.**

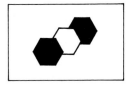
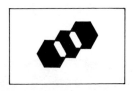

Specification

Pronounced: SPESI-FIK-A-SHUN (*e as in let, i's as in ink, a as in late, u as in fun*)

Origin

From the Latin *species* meaning *kind* or *sort* and *facere* meaning *to make* (compare with **fac**tory). *Specify* came to mean *to make mention of specific kinds of things - that is definite requirements for work to be undertaken.*

Meaning

A specification refers to the written documentation (including working drawings and blueprints) which states the materials, the quantities, the forms and methods of construction or manufacture required and the costs involved to ensure that the problems identified in a **design brief** are solved. A specification is only one of the elements in the production of a designed object, which requires:

- consultation between a designer (or design team) and a client (the *project initiation*), where problems are identified;
- a project management schedule, where the design work to be undertaken to complete the *project* is stated, and a time schedule for the work and a schedule of fees are agreed to;
- a design brief, investigating and analysing relevant data and identifying the *design criteria* by which the design problems can be solved;
- a specification;
- the design work, incorporating all aspects mentioned above;
- production of the designed object.

Associations

See: **design, drawing**.

Spectrum

Pronounced: SPEK-TRUM (*e as in let, u as in drum*)

Origin

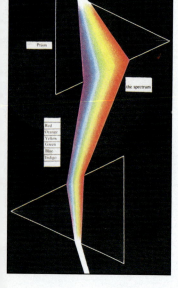

Spectrum is the Latin for *image, ghost, spectre, phantom*. The term was used into the 17th century for an *apparition* and then was used more for an image than a ghost towards the end of the 17th century. Sir Isaac Newton used the word spectrum in 1671 in his work on the composition of light.

Meaning

The spectrum is the coloured band into which a beam of white light appears after it has been passed through a **prism** or other source which diffracts (*breaks into pieces*) the light (e.g. a diamond, a water droplet or a soap bubble). The prismatic hues of the spectrum are: red, orange, yellow, green, blue, indigo, violet.

Each coloured light has a different wave-length. Violet has the shortest and is bent most by the prism; red has the longest and is bent the least.

Associations

Iridescent means displaying the prismatic colours of the rainbow, or flashing with changing colours.

See: **colour, colour-wheel, facet, filter, pigments**.

Spline

Pronounced: SPLIN (*i as in fine*)

Origin

From an Old English word meaning *a splinter*, a thin piece of wood or metal, usually used as a small wedge.

Meaning

In metalwork, a spline is a *key*, usually rectangular, which fits into a groove in a hub or shaft. It allows some longitudinal play. The metal key and the groove receiving the key are called a *feather*.

Associations

See: **clearance, groove, joint, key, line.**

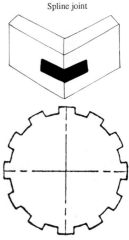

Spline joint

end view of a splined shaft

Squaring Up

Pronounced: SKWAR-ING UP (*a as in fare, u as in cup*)

Origin

From the Latin *exquadrare* meaning *to square (make four sides and four angles)* and *quatuor* meaning *four*. Squaring up was used in wall paintings in Ancient Egypt and may have been used even earlier.

Meaning

Squaring up is a method of transferring a drawing on one scale to a larger scale. The technique is often used where a drawing has to be transferred to a large area such as a large canvas or piece of plywood or to a wall. The technique is to cover the drawing with a grid of numbered squares. The same number of larger squares is placed, as a **grid**, on the area which will take the enlargement. Each square of the drawing is then copied onto the corresponding square of the larger surface.

Associations

Squaring up is also called **graticulation**.
See: **grid, pantograph, proportion, replica, scale.**

stake and holder

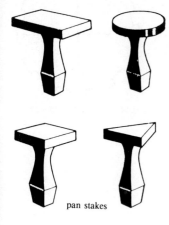

pan stakes

Stake

Pronounced: STAK (*a as in make*)

Origin

From Old English *stacca* meaning *a sharpened stick driven into the ground as a support.*

Meaning

A stake is a metal worker's steel tool, which is fixed (originally by a pointed end) onto a bench, using a stake holder or into an anvil or a hardwood block. It provides a hard working surface to support sheet metal while it is being shaped with a hammer or mallet. There are numerous shapes of stakes, which are variations of three basic shapes: anvil, mushroom and T-shape. As the sheet metal is placed over the stake and then worked, the metal assumes the shape of the surface of the stake. Any shaped piece of material, such as wood, plastic or metal which is used to shape sheet metal is now called a *stake*.

A stake is also a piece of timber which is pointed at one end to make it easy to drive it into the ground.

Associations

See: **burr, dapping, dress, hammer, repoussé, shape**.

Stamp

Pronounced: STAMP (*a as in cat*)

Origin

From the Old English *stampian* meaning *to bring down one's foot heavily* and later *to press a mark into something.*

Meaning

To stamp is the process of pressing or embedding an object into a material in order to leave an impression of the object on the material. In **block printing** this is simply cutting a shape on a block (of wood, lino or rubber etc.), inking the block and then pressing the block onto paper or fabric.

Patterns can be made on metal by stamping with metal punches. A stamp or **die** can be used to stamp on leather and stamps can be carved in plaster (with the

design in reverse) to make patterns on potteryware.
Associations
See: **die, emboss, impression, pattern, swage, symbol.**

Stencil

Pronounced: STEN-SAL (*e as in pen, a as in ago*)

Origin

From the Middle English *stanselen* meaning *to ornament with sparkling colours or pieces of metal*. The word probably changed from the product to the process to get the product. The device of a stencil originated in China in the 6th century. The stencil was held together by fine strands of hair or silk. Stencils were used to make playing cards in the 15th century and for wallpaper in the 17th century. They were used to decorate walls until the 19th century when the use of wallpaper became a reasonably cheap way of decorating walls. As early as 1775, stencils were made on furniture. They were used extensively in the United States during the 1800's for decorating floor coverings, walls and household fabrics.

Meaning

A stencil is a thin, re-usable sheet of metal, cardboard, paper or plastic in which lettering or a design is cut out. The stencil is placed on paper, card, canvas, wood, metal, fabric, vitreous enamel objects, or other materials and the cut-out parts are painted, shaded, inked or dyed. These areas transfer to the material under the stencil to reproduce the lettering or design on the stencil. The uncut areas act as masks. Stencils are usually used where a repetitive design is needed.

Associations

A stencil is sometimes called a **mask**, because it masks areas of a composition not to be reproduced.
See: **enamel, development, motif, pattern, template.**

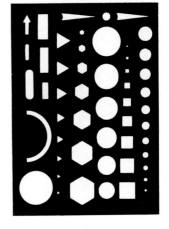

Stile

Pronounced: STIL (*i as in file*)

Origin

From the Old English *stigel* meaning a *step*, or *set of steps* for climbing over a wall or fence.

Meaning

A Stile is the vertical (upright) part (called a *member*) of a wooden frame. The horizonal member is called a **rail**.

Associations

See: **carcass, furniture.**

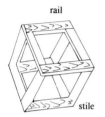

Stop

Pronounced: STOP (*o as in lot*)

Origin

From the Old English *stoppian* meaning *to prevent or forbid passage of something or somebody*

Meaning

A **stop rod** is an adjustable rod which can be set to ensure a number of pieces of material (e.g timber) are cut to the same size. A **bench stop** is an adjustable piece of timber or metal fitted to a bench to prevent a workpiece from moving when it is being planed.

Stopping is a compound (e.g. putty or beaumontage) which is used to fill holes (such as borer holes, nail holes) and cracks in wood before a finishing process begins. The stopping is usually coloured to match the colour of the wood.

A **bit stop** is a simple device which is fitted to a bit to control the depth to which the bit drills.

In photography, a stop is a device for changing the diameter of a lens to control exposure time.

Associations

See: **bit, drill, exposure, finish, lens, plane.**

a stop block

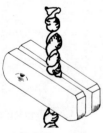

a depth stop for boring

Swage

Pronounced: SWAJ (*a as in late*)

Origin

From Old French *souage* meaning *a decorative groove*

Meaning

One meaning of *to swage* is to place beads or bands of metal on sheetmetal objects (e.g. a drum or a bin) to increase their strength. Another is to alter the shape of metal by any means (e.g. by hammering, bending, rolling) other than by cutting, sometimes so that pieces of sheetmetal may be joined.

A hole is often swaged in sheetmetal to assist tapping a thread into it or to obtain a thick rounded edge to a hole to allow the smooth passage of anything (e.g. wires or ropes) passing through the hole.

A swage is a die or stamp for shaping metal (e.g. wrought iron) by hammering or by some other means of pressure. It is also a tool for bending metals and a two-piece tool used for shaping hot metal rods. A **swage block** is a cast iron or steel block which has patterned edges and surfaces with various-shaped holes. Hot or cold sheet metal placed over the patterned parts and then punched or hammered will receive the impression of the patterns.

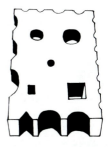

a swage block

Associations

See: **bead, burr, dapping, die, emboss, hammer, impression, pattern, stamp, shape.**

Swarf

Pronounced: SWORF (*o as in lord*)

Origin

From an Old Norwegian word *svarf* meaning *file-dust*, the fine chips or grit from a grindstone or filings from metal

Meaning

Swarf is the waste metal which is produced when machining metal.

Associations

See: **machine.**

Symbol

Pronounced: SIM-BAL (*i* as in *duty*, *a* as in *ago*)

Origin

From the Greek *sumbolon* meaning *mark* or *token*. Symbols are ancient methods of expression by all people.

Meaning

A symbol is an object or drawing which represents or signifies some idea or quality by association in fact or thought, and which gives an instructional message. For example, a white dove is a symbol of purity, love, peace or the Holy Spirit; the lion symbolises courage, the cross Christianity, a heart love, an arrow a direction, a halo holiness. The International Travel Communication Signs and symbols in electrical systems are good examples of the importance of symbols in our world. Symbols are used in almost every specialised area, for example in weather signs, mathematics, maritime affairs, graphics, welding. A well-designed commercial or industrial symbol (or logo) should be simple and easily identified, easy to produce and photograph in black and white, and should be easily remembered.

toxic (poisonous)

explosive

risk of electric shock

oxidizing

laser radiation

harmful

corrosive

non-ionising radiation

highly flammable

safety code symbols

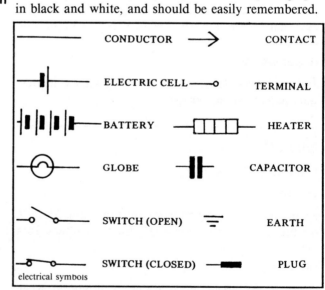
electrical symbols

Colours also had symbolic meanings in the past, and to a lesser extent today, such as white for purity, red for love, green for hope, blue for truth, and yellow for treachery.

236

Associations

A **logo** is a distinctive, designed image using one or more letters, which represents or symbolises an organisation. **Pictographs** are pictures as symbols or signs. An **ideogram** is a letter or character which symbolises an idea.

See: **emboss, hallmark, impression, stamp, template**.

Symmetry

Pronounced: SIM-A-TRY (*i as in limb, a as in ago, y as in duty*)

Origin

From the Greek *syn* meaning *together* and *metron* meaning *measure*.

Meaning

Symmetry is a situation where one part of something (e.g. the human body) is of the same measure or proportion as another part and a harmonious balance of parts results. Usually objects or structures (e.g. the columns of an arch) are on different sides of a dividing line (called *lines of symmetry* in graphics) and the sizes, shapes, colours and textures of the objects on both sides are the same and balanced. When symmetry occurs the objects are said to be **symmetrical**

Associations

Asymmetry is the opposite of symmetry. It is when there is no balance or harmony of the objects on the two sides of a dividing line.

See: **distortion, form, geometric, pattern, proportion, replica, scale, shape**.

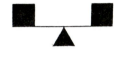
symmetry

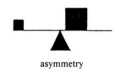
asymmetry

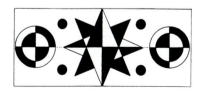

Synthetic

Pronounced: SIN-THETIK (*i's as in pin, e as in pet*)

Origin

From the Greek *sunthetikos* meaning *a placing together, combining*. The production of synthetic materials from chemicals is a 20th century development and the impetus to produce such materials gained momentum during and after the 1939-1945 war, when natural resources were at a low ebb. Plastics particularly (which are synthetic resins) were found to be a satisfactory substitute for wood, ceramics, glass, rubber and other substances. They were materials which could be readily processed and mass produced and could often be shaped into forms which was not possible with natural substances. The development of synthetic adhesives and synthetic finishers since the 1940s and 1950s has greatly affected the production of a wide range of materials in most industries

Meaning

Synthetic is the adjective from **synthesis**, which means the putting together or building up of separate elements into a connected whole. It usually implies that relatively simple elements have been **synthesised** (combined) to form a complex composition.

For example, when carbon fibre, consisting of long molecular chains of carbon as reinforcing rods, is laminated inside sheets of synthetic resin, the resultant material has twice the strength of steel yet only a quarter of its weight.

Associations

The opposite of synthesis is **analysis** (from the Greek meaning *loosening up*), which means the breaking up of the whole into its elements.

See: **adhesive, cross-linking, finish, lacquer, plastic, plywood, resin, thermoplastic, thermosetting, vulcanise.**

Tap

Pronounced: TAP (*a as in map*)

Origin

From the Old English *taeppa* meaning *a tapering stick*, which was used as a plug for stopping a hole in a cask of liquid. *To tap* then came to mean to remove the tap to draw off the liquid in the cask. Tap also came to mean the device (e.g the pipe) by which the liquid was drawn off in a controlled way.

Meaning

To tap is either the process of making internal (female) threads in a cylindrical metal object, or the process of drawing off molten metal from a furnace.

A *tap* is a fluted, threaded tool for cutting or forming internal (*female*) threads. Some taps cut a thread; others do not break the metal grain but squeeze and form the metal (usually soft metals, such as brass or aluminium) which produces a strong thread. There are usually three different taps for each size of thread: the **taper tap** with between 6 and 9 of its end threads which taper to a wedged end. This is used to grip the metal and to start the internal thread; the **second** or **intermediate tap** with 3 to 6 end threads which taper, which is used to follow the taper tap; the **plug tap** (or **parallel tap**), which is not tapered and which is used to reach the bottom of the thread hole. It is never used to start a thread. A **tap wrench** is a hand tool which is used to hold a tap. It has long handles on each of its sides to provide sufficient leverage to turn a tap in a thread hole.

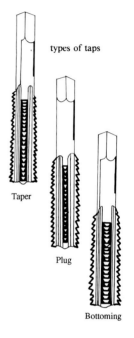

types of taps

Taper

Plug

Bottoming

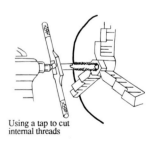

Using a tap to cut internal threads

Associations

A **tap bolt** is a bolt with a head at one end and a thread on the other end. It is screwed into some fixed part instead of passing through the part and receiving a nut.
See: **bolt, fastener, screw, taper, threads**.

Taper

Pronounced: TAPA (*1st a as in late, 2nd a as in ago*)

Origin

From the Old English *tapor* meaning a *wax candle*, which was long, thin and gradually became narrower

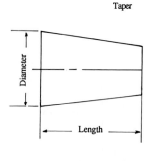
Taper

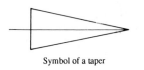
Symbol of a taper

and rather cone-shaped at the top.

Meaning

An object is said to have a a taper when it decreases uniformly in width or diameter along its length to make a wedge or conical shapes (for example, like some table legs, a wedge or a lathe-centre).

Tapers are used in workshops to ensure that there is perfect fit and grip in some tools and machines. For example, the spindles of some machines (e.g. a drill press or a milling machine) have internal tapers, which fit tightly into the external tapered parts of the machines they are made for.

Associations

A **tang** is a tongue machined on the end of a tapered shank, which fits into a slot in the mating part to stop the taper from turning in the mating part. A **taper gauge** is used to test the accuracy of both inside and outside tapers.

See: **draft, ferrule, fit, gauge, saw, scarfing, tap.**

Technical Drawing

Pronounced:TEK-NI-KAL (*e as in pen, i as in ink, a as in ago*)
See: **draw**

Origin

Technical is from the Greek *tekhne* meaning *art*, which implied *skill* and *craft*. The term *art* was used to describe skill in a wide range of crafts and professions with no emphasis on specialised areas until the 17th century. The use of *art* to apply *only* to painting, sculpture and other acts of creative expression did not become established until the 19th century

Technique is from the Greek *tekhnikos* meaning *art* or *craft* and *made by art*

See: **drawing** and **projection**..

Meaning

Mechanical drawing has many of the features of technical drawing (e.g. projections, dimensioning, rendering, etc.) but mechanical drawing relates closer to industry and to the production of **assembly drawings** and **parts drawing** of machines. Mechanical drawing

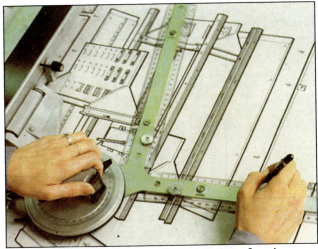

usually involves numerous separate drawing of various parts of a machine, whereas a technical drawing can usually be depicted on a single sheet.

Associations

See: **angle, cross-section, cross-hatching, develop, draft, drawing, format, geometric, Golden Section, grid, industrial design, line, perspective, projection, sketch, template, trammel**

Temper

Pronounced: TEMPA (*e as in let, a as in ago*)

Origin

From the Latin *temperare* meaning *to mix in due proportion, to mingle, to moderate, to regulate.*

Meaning

The temper of a metal is its correct degree of hardness. *To temper* metal is to heat it to a specified temperature (being the lowest temperature that will give a particular metal its required hardness - usually between 200°C to 300°C) for a specified time and then to cool it quickly, either by blasting it with cold air or by dipping it in water, brine or oil - a process called **quenching**. The temperature of the liquids used for quenching vary considerably and must be accurate for different metals to be effective. The metal is under stress during the process and may crack if it is cooled too rapidly.

Heat and Temper Colours

COLOUR in dull light	APPROX TEMP °C	°F	COLOUR in dull light	APPROX TEMP °C	°F
Brown Red	565	1049	Faint Straw	205	400
Dull Red	680	1256	Straw	225	440
Blood Red	730	1346	Deep Straw	245	475
Medium Cherry	750	1382	Bronze	270	520
Cherry Red	780	1436	Purple	280	540
Bright Cherry	825	1517	Full Blue	295	563
Full Red	850	1562	Light Blue	310	590
Yellow Red	950	1742	Grey	330	626
Orange	1050	1922			
White	1300	2372			

The tempering and drawing temperature differs from one steel to another. The recommended times and temperatures for various steels are supplied in a steel manufacturer's handbook.

Tempering lessens brittleness in metals and helps to toughen them.

The temperature of steel can be judged by the colour of the polished steel while it is being tempered, ranging from pale yellow at 430°F through brown at 490°F to dark blue at 600°F. In photography, a **tempering box** is a control-unit which holds bottles and **graduates** of solutions, which, by means of an electrical heater, a thermostat and an air or water circulatory system, can be kept at a required temperature. It may hold the film tank.

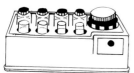
Thermostatically-controlled tempering box

Associations

Tempering is also called **drawing**.
See: **annealing, burin, hardening, metallurgy**.

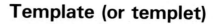

Template (or templet)

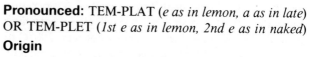

Pronounced: TEM-PLAT (*e as in lemon, a as in late*) OR TEM-PLET (*1st e as in lemon, 2nd e as in naked*)

Origin

From the Latin *templum* meaning *an open space marked out for observation of the sky*. The word *tem* became a *marked space* and when joined to *plate*, meaning a *flat shape*, a template became a flat device used for *marking out* things, usually those which needed to be repeated accurately in exactly the same form.

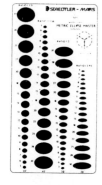

Meaning

A template is a pattern or gauge, usually of thin sheet-metal, plastic, plywood or cardboard, which is used as a guide to ensure accurate drawing, cutting, drilling or shaping of metal, wood, plastic, stone, clay, leather, etc. It is frequently used where a pattern or profile has to be repeated using standard shapes or symbols.

A **template former** is a tool for copying the profile of shapes.

Associations

See: **development, gauge, mechanical drawing, pattern, profile, shape, stencil, symbol, technical drawing.**

marking out from a set-out rod

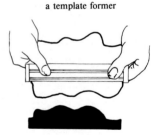

a template former

Tensile

Pronounced: TEN-SIL (*e as in men, i as in file*)

Origin

From Latin *tensibilis* meaning *that which may be stretched*. Tension originally meant *the act of stretching* or the *state of being stretched*
Leonardo da Vinci (1452 - 1519) invented a machine to test the tensile strength of wire. Robert Hookes (1635 - 1703) tested the tensile strength of long wire and formulated the law of *linear elasticity*. Musschenbroek invented a tensile testing bending machine. Today non-destructive testing of tensility can be carried out by X-ray methods.

Meaning

Tensile describes metal which is able to be drawn or stretched without breaking. The **tensile strength** or **tensility** of a metal is the maximum stretching load it can bear *under tension* before it breaks.

Associations

See: **drawing, ductile, fatigue, hardening.**

Texture

Pronounced: TEX-CHA (*e as in let, a as in ago*)

Origin

From the Latin *textura* meaning *weaving*. Originally the word applied to the feel of a woven garment and then was used for the feel of any material or surface.

Meaning

Texture refers to the nature of the surface of a work (e.g. matt, gloss, smooth, grained, corrugated, etc.). A textured effect can be **tactile** and felt with one's hand (e.g the roughness of tree bark).

The texture of wood is determined by the size of the pores (or **vessels**) of the wood. The smaller the pores, the finer the texture of the wood. Meranti, for example, has large pores and has a coarse texture.

In photography, texture refers to the quality of the surface of a photograph.

Associations

See: **bromide, contour, elements, grain, marquetry, matt, pored.**

Thermodynamics

Pronounced: THURMO-DI-NAMIKS (*u as in fur, o as in go, i as in line, a as in plan, final i as in ink*)

Origin

Thermo is from the Greek *therme* meaning *heat*; *dynamics* is from the Greek *dunamikos* meaning *power*. The word *dynamics* was introduced into the English language by Jeremy Bentham (1748 - 1832), an English political philosopher. *Thermodynamics* was first used by an English physicist, James Prescott Joule (1866 - 1889), to explain the relationship of heat to work and the mechanical equivalent of heat. James Clerk Maxwell (1831 - 1879), a Scottish scientist, made an outstanding contribution to the understanding of thermodynamics.

Meaning

Thermodynamics is the branch of physics which deals with heat as an agent of mechanical power. The first law of thermodynamics states that: "*when work is transformed into heat, or heat into work, the quantity of work is mechanically equivalent to the quantity of heat*". The second law of thermodynamics states that: *the heat tends to flow from a body of hotter temperature to one that is colder and will not naturally flow in any other way*.
The science of thermodynamics deals with concepts such as temperature, energy, kinetic theory and the relationship between heat and work.

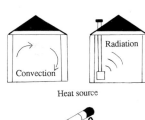

Heat source

Three primary methods of heat transfer

Associations

See: **kiln, pyrometry**.

Thermoplastic

Pronounced: THURMO-PLASTIK (*u as in fur, o as in go, a as in pass, i as in ink*)

Origin

Thermo is from the Greek *therme* meaning *heat*; plastic is from the Greek *plastikos* meaning *that can be moulded or shaped* from *plassein* meaning *to mould or shape*. Thermoplastic polystyrene was discovered in the 1830 but was not commercially produced (in Germany) until about 1930. Polypropylene, now one of the most commonly-used thermoplastics, was first produced in a useable form by Professor Nalta of Italy in 1954. Polycarbonates, a versatile group of thermoplastics, was first commercially produced in 1960.

Meaning

A thermoplastic is a synthetic material which can be made pliable by heating it. While it is hot, it can be shaped as required. It will keep this shape as it cools and sets. That is, it has what is termed **dimensional stability**. It can be reheated and reshaped several times, in contrast to a **thermosetting** material which once it has been heated and shaped cannot be reheated to be reshaped.

Some commonly-used thermoplastics are:**polystyrene**, which is used for food containers, ceiling tiles and packing for delicate articles; **cellulose acetate**, which is used for spectacle frames, toothbrush handles and transparent packaging film; **acrylic**, which is best known as a glass substitute, under the trade name *perspex.*

Thermoplastic materials can be strengthened and reinforced by the addition of such materials as glass, cotton, nylon fibre, carbon fibres and asbestos. Some thermoplastics which are reinforced by fibre-glass are nylon, P.V.C. and polypropylene. Thermoplastics are used extensively when an object is to be coated in plastic.

Associations

A **thermo-former** is a machine for heating and forming thermoplastic sheets.

See: **blow moulding, calender, cross-linking, curing, extrusion, injection moulding, moulding, plastic, resin, synthetic, thermosetting, vacuum forming, weld**

Thermosetting

Pronounced: THURMO-SET-ING (*u as in fur, o as in go, e as in let*)

Origin

Thermo is from the Greek *therme* meaning *heat. Set* is from the Old English *settan* meaning *to sit* or *to put a thing in its place.* From these meanings developed the meaning of *to fix* or *to make something immoveable.* Epoxy resins (thermosetting plastics) were known in the late 1800s but they were not developed until about 1930. The first thermosetting material to be formed entirely from chemicals in a laboratory was phenol formaldehyde. It was made by Leo Baekeland in 1907, and although its industrial name was *phenolic*, it became known as *bakelite*.

Meaning

A thermosetting plastic material can be moulded into infinite shapes when it is heated but after it has been cured it cannot be made soft again or reshaped (See: **cross-linking**). **Urea formaldehyde** is one of the oldest known forms of thermoset resins. **Epoxy** is a thermosetting resin which cures to a hard surface. It has good adhesive qualities and is resistant to most chemicals. **Melamine (melamine formaldehyde) is a thermosetting powder used to make** *laminex* for good-quality table ware. Like thermoplastic materials, thermosetting materials can be reinforced by the addition of other materials. The thermoplastics most commonly used for *fibre-reinforced plastic* (F.R.P.) are polyester and epoxy.

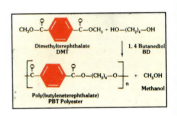

Associations

Post-forming is a process used to bend or shape a hardened thermosetting laminate.

See: **cross-linking, curing, injection moulding, laminate, plastic, resin, thermoplastic, weld.**

Thicknesser

Pronounced: THIK-NESA (*i as in ink, e as in let, a as in ago*)

feeding stock through a thicknesser

247

Origin

Thick is from the Old English *thicce* meaning *dense and to a great depth*.

Meaning

A thicknesser is a power-driven machine which is used for planing timber (**stock**) of more than 300 mm in length (the minimum for safety reasons) to a uniform thickness, which can be measured by a calibrated vernier scale attached to the machine.

Any defects in the stock (such as cupping or winding) are first removed and the timber is made straight, square and true on two surfaces on a **planer jointer** before the thicknesser is used.

Associations

See: **distortion, plane, scale**.

Third Angle

Pronounced: THURD - ANGAL (*u as in fur, 1st a as in ant, 2nd a as in ago*)

Meaning

When horizontal and vertical planes intersect at right angles , four angles of projection may be visualised. They are known as **dihedral angles**. To provide essential information on a technical or mechanical drawing,

three views (of the total six available) from an angle of projection are needed, namely a front view, a top view and a left side view. This information gives an object's length, breadth and depth and an impression of its shape. This type of drawing is called an **orthographic** or **orthogonal projection**, or, more commonly, a **working drawing**. Most working drawings in the past have used a First or Third Angle projection or a combination of the two, but nowadays most drawings use a Third Angle projection where the object is placed so that it shows the side of the object nearest to it on the adjacent view.

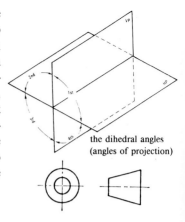

the dihedral angles (angles of projection)

symbol of third angle projection.

Associations

See: **mechanical drawing, projection, technical drawing**.

Threads

Pronounced: THREDZ (*e as in red*)

Origin

From the Old English *thraed* meaning *something which is thrown and twisted* from the verb *thrawan* meaning *to throw, to twist*.

See **screw**.

Meaning

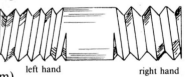

left hand   right hand

Threads are the promiment, twisted (in spiral form) ridges below the shank of a screw. Threads can be right-handed or left-handed. A right-handed thread is one which advances when moved in a clock-wise direction; a left-handed thread advances when moved in an anti-clock-wise direction. Most threads are cut right-handed. A thread can be cut by a **tap**, which is a sharp, fluted, threaded cutting tool. **Tapping** is the threading process, which can be made either by hand (using a **tap wrench**) or by the use of a machine.

Over the years, there have been a variety of thread forms produced, each with its own characteristics. There are six common thread systems in use throughout the world today. They are: British Standard Fine (B.S.F.), British Standard Whitworth (B.S.W.), Unified National Fine (U.N.F.), National Fine (N.F. or S.A.E.), National Coarse (N.C.) and Metric .

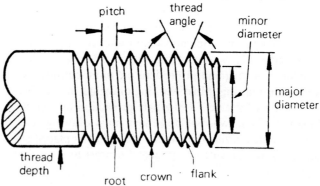

parts of a screw thread

Associations

The surface which lies between adjacent threads is called the **root** at the bottom and the **crest** or **crown** at the top.

See: **gauge, pitch, screw, tap.**

Straightness	—
Flatness	▱
Roundness	○
Cylindricity	⌭
Profile of a line	⌒
Profile of a surface	⌓
Parallelism	∥
Squareness	⊥
Angularity	∠
Position	⌖
Concentricity	◎
Symmetry	⌯

Tolerance symbols

Tolerance

Pronounced: TOLA-RANS (*o as in box, a's as in ago*)

Origin

From the Latin *tolerare* meaning *to bear, support, endure.*

Meaning

Tolerance in machining is the precise and agreed variation in the sizes of machined parts. It is the limit of variation that is generally acceptable in a given situation. **Unilateral tolerance** is a variation from a basic dimension which is allowed one way, plus or minus but not both ways. **Bilateral tolerance** is a stated tolerance which is both smaller or bigger (+ or -) than a specified , basic dimension, e.g 15.25 ± 0.05. The degree of accuracy required in work depends upon the type of work undertaken, and the degree of tolerance generally accepted varies from one type of work to another.

Metrology is a branch of technology which specifies the tolerance with which mechanical parts should be made in a machine-shop or production line.

Associations

See: **callipers, clearance, drawing, fit, gauge, machine**

Tone

Pronounced: TON (*o as in alone*)

Origin

From the Greek *tonos* meaning *tension* and *stretching*. From this came the meaning of stretching in the sense of a gradation, so different tones showed a range of different intensities.

Meaning

To tone is to change the contrasts of a black and white photograph. **Tone separation** is where a photograph has its normal range of black, greys and white made into two or three flat tones between black and white. **Tone drop out** describes a photograph from which all grey has been removed leaving only black and white.

By a chemical process (e.g. by using sodium sulfide) called **toning** (often combined with **bleaching**), any black and white photograph can change its black silver image into another colour. Sepia toning, which produces a rich chocolate colour (sepia) is a popular method of toning.

A **tone-line** is a process in photography where a continuous-tone image is converted into a white or black contour image which is very similar to a pen and ink drawing

Associations

See: **contrast, gradation.**

Tooling

Pronounced: TOO-LING (*oo as in soon*)

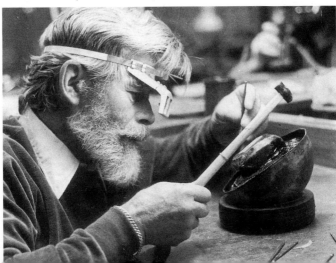

Origin

From the Old English *tol* meaning *a tool*.

Meaning

Tooling is the smoothing (also called **dressing**) of stonework, using a chisel. It also means to decorate metalwork, either in relief or intaglio, using tools such as punches, gravers and hammers. This is also called **chasing**. *To tool* also means to decorate leather books with designs, using heated tools which burn into the leather.

Associations

See: **chase, dress, intaglio, relief, sculpture**.

Torque

Pronounced; TORK (*o as in bore*)

Origin

From the Latin *torques* meaning *a necklace made of twisted metal*. It originated from *torquere* meaning *to twist and bend*. The word came to mean something which twists and rotates, such as a spindle, and then a twisting force.
Charles Coulomb (1736 - 1806), a French physicist, invented a torsion balance which measured torque by the amount of twist produced in suspended fibres,

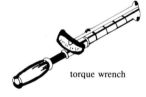
torque wrench

Meaning

Torque refers to the forces which produce a twisting or rotating motion. The turning effect of a belt on a pulley is called torque

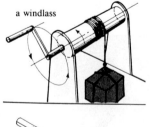
a windlass

Associations

Torsion means spirally twisting and the tension which is exerted on a metal bar or wire when one end is fixed and the other is rotated about its axis. A **torque wrench** is a special wrench which has a scaled indicator which shows the amount of force applied to the object being turned and tightened.

a crank

A **crank** is a device by which torque can be applied to a shaft (e.g. on a windlass).
See: **scale, wrench**.

Trammel

Pronounced: TRAMAL (*1st a as in bat, 2nd a as in ago*)

Origin

From Latin *tramaculum* meaning a *triple drag-net* for catching fish. It then came to mean anything which restricts movement. The development of the term is uncertain but its meaning as a tool probably originates from the tool's controlled and restricted range of operation.

Meaning

A trammel is a tool used to mark out distances, large circles and arcs on a material (such as wood, paper, plastic or sheetmetal), which are beyond the range of dividers and winged compasses. It comprises a bar or beam (a straight edge) which passes through two or more *trammel heads,* each of which is fitted with an instrument which will record the shape required (e.g. a scribing point) and which can be moved along the bar and adjusted by means of milled-edged set-screws to produce the size of circle etc. required. A trammel frame with three scribing points can be used for setting out an ellipse.

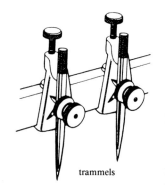
trammels

Trammels are also known as **beam compasses.**

Associations

See: **drawing, mechanical drawing, technical drawing**.

Transistor

Pronounced: TRAN-ZISTA (*1st a as in band, i as in pin, final a as in ago*)

Origin

Transistor is a word coined during this century. It is a portmanteau word (a word combining the meaning of two or more other words) from *transfer* and *resistor*. The word was first used by the Bell Laboratories in the United States in 1948, when the transistor was invented.

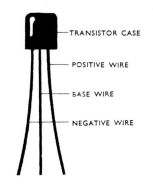

Meaning

A transistor is a semicondcuctor component with three or more electrodes which is used for switching and amplifying. It is made of small pieces of pure silicon.

It performs the same function as a thermionic valve (an incandescent valve that emits electrons in one direction) but is smaller, stronger and consumes less power. It is a component, as small as a pin prick, in an integrated circuit and has been important in the miniaturisation of electronic components.

Associations

See: **circuit, conductor, electrode.**

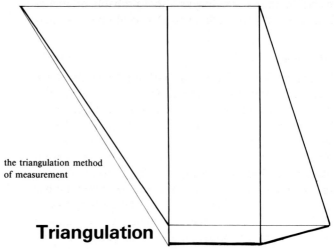

the triangulation method of measurement

Triangulation

Pronounced: TRI-ANG-U-LASHUN (*i as in fine, 1st a as in cat, u as in due, 2nd a as in late, u as in bonus*)

Origin

From the Latin *triangulum* meaning *a triangle* and *angulare* meaning *to make angular.*

Meaning

Triangulation is a method of measuring the areas of irregular uncurved shapes by dividing the shapes into triangles, calculating the area of each triangle and then adding the areas of all the triangles making up the shape. Triangulation is the most important method of *development* (See: **Drawing**) used in industry, which is often concerned with products with complex shapes and with changing square pieces of material into circular shapes.

Triangulation is also known in surveying as **intersection**. **Radial line development** is a form of triangulation.

Associations

See: **development, drawing, shape.**

Turn

Pronounced: TURN (*u as in fur*)

Origin

From the Latin *tornare* meaning *to turn on a lathe*. The Latin word *tornus* meant a *lathe* and the Ancient Greek word *tornos* meant a *tool for drawing circles*. See: **lathe**.

Meaning

Plain or parallel **turning** is a process on a turning-lathe where an inclined cutting tool is fed along a work parallel to its axis usually to produce a cylinder-shaped object.

Turnery is the art of turning, things made by a **turner** and a turnery is the place where turned objects are made.

In recent years, wood-turners have begun to work with a variety of materials, such as plexiglas, DuPonts Corian and rotting or unseasoned timber.

Associations

See: **arbor, chisel, face, lathe, machine, mandrel**.

turned objects

Vacuum forming

Pronounced: VAK-U-UM (*a as in bat, 1st u as in due, 2nd u as in bonus*), FORM-ING (*o as in born*)

Origin

From the Latin *vacuum* meaning *an empty space*.

Meaning

A vacuum is a space which is entirely without matter. Usually air has been removed by a pump from the space. A perfect vacuum, one that contains no gas at all, is unobtainable. The pressure in the vacuum is substantially below normal atmospheric pressure.

Vacuum forming is a process in the forming of plastic objects which makes use of the fact that normal atmospheric air can exert pressure sufficient to move material which is situated in a vacuum.

Thermoplastic in sheets is heated until it is plastic. The sheets are then placed over a mould in a vacuum. When air pressure is placed on the plastic material, it is blown down onto the mould, where it is cooled and formed into the shape of the mould.

Associations

See: **blow moulding, moulding, plastic, thermoplastic.**

a vacuum-forming process

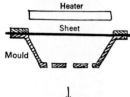
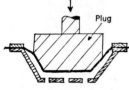
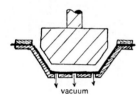

Varnish

Pronounced: VAR-NISH (*a as in car, i as in dish*)

Origin

From the Middle English *vernisch*, which is from the Latin *veronic* meaning *a fragrant resin*, which was probably sandarac. Varnish was first exported from what is now Benghazi in Libya. Resins and their use to make varnish have been known for hundreds of years. Recipes are given in some early Medieval documents. *Oil varnishes* were common from the 9th to 15th centuries (using mastic, sandarac and linseed oil) and *spirit varnishes* were used from the 16th century. *Synthetic resins* are a product of the 20th century.

Meaning

A varnish is a liquid which is used to give a protective coat to wood, metal and other materials. It dries hard and is usually glossy and transparent, although nowadays there are matt varieties. It is able to protect objects from damp, pollution, grease and corrosive substances. It is easily applied (preferably in warm, dust-free environments), using a brush or spray, and it is readily removed, if necessary, with a **solvent**.

There are two kinds of varnish. One is **oil varnish**, which is made from a combination of a hard, fossil resin (such as copal made from amber) heated with a drying oil, such as linseed oil, and then thinned with a solvent. The other is **spirit varnish**, which is made from a soft, natural resin (such as mastic, sandrac or dammar), which is dissolved in alcohol. Natural resins are secretions from some plants and trees (See: **lacquer**). These resins are adhesive, inflammable and insoluble in water. Natural resin varnishes tend to turn yellow with age and become brittle. Consequently, **synthetic varnishes** have been produced which do not have these faults. These consist of synthetic resins (such as bakelite, alkyds and polyurethane) and China Wood oil, with xyol or varnalen as a thinner.

Varnishes are used extensively not only as a **fixative** or a **resist agent** but also to provide a gloss or matt glaze to a finished work. **Spar varnish** (so called because of its use on the spars of ships) is a very durable varnish which is resistant to sunlight, heat and rain and is most suitable for use outdoors.

Associations

Shellac is a varnish used often on timber. Shellac dissolved in methylated spirits is also known as **French polish**. **Crawling** is a defect in varnishing where there is poor adhesion of varnish to a surface in some places and the varnish "crawls" into globules of varnish instead of being spread out smoothly.

See: **accelerator, glaze, lacquer, matt, preservation, priming, resin, resist, solvent, vehicle, viscosity.**

Vehicle

Pronounced: VEE-I-KAL (*ee as in see, i as in pin, a as in ago*)

Origin

From the French *vehicule* from the Latin *vehere* meaning *to carry* and *vehiculum* meaning *a carriage* or *a means of transport*.

Meaning

A vehicle is the **medium** or **binder** which holds the ingredients in a liquid in suspension . Each vehicle (e.g. oils, resins, latex, cellulose, bitumens and varnishes) has distinct properties which it brings to the liquid. It usually is the agent for forming a hard surface film on an object after drying, provides adhesion and makes a surface durable and washable. In addition, it can have other functions. For example, the function of a vehicle in paint is to make pigments into a liquid or a paste form, so that the mixture can be spread (that is "*carried*"). For example, linseed oil or tung oil is a vehicle for oil paint or oil varnish, varnish for varnish enamel and acrylic emulsion for acrylic paint.

Solvents dissolve or disperse a vehicle, but do not remain as part of the surface film as they evaporate.

Associations

See: **enamel, finish, paint, pigment, solvent varnish**.

Veneer

Pronounced: VA-NEAR (*a as in ago, ea as in fear*)

Origin

Veneer used to be called *fineer* from the German *furnieren* meaning *to furnish*.

The use of veneers goes back more than 3500 years. Records show the use of the technique in Ancient Egypt, Greece and Rome. Veneering was practised extensively in Europe in the 17th century and was introduced into England in the latter part of the 17th century. In 1830, a process of steaming and slicing flitches of timber for thin veneers was invented in Paris. See **marquetry**.

Meaning

A veneer is a thin strip of wood (often figured and expensive) or other material (e.g. ivory) which is overlaid or faced on another (usually inferior) piece of wood for decorative purposes. The veneers vary in thickness from 0.25 mm to 1.2 mm. They are produced by lathes, slicers or saws and are sliced or, more generally, rotary peeled from timber.

Particle-board, hardboard, plywood and blockboard are

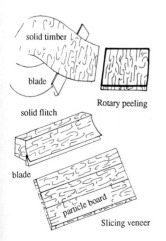
Rotary peeling
Slicing veneer

often used as the bases on which veneers are overlaid. In industry, a **veneer press** (usually either a **vacuum press** or a **dome press**) is used to adhere veneers to bases. In schools, the *caul and clamp* method is usually used, with P.V.A. as an adhesive.

A veneer can also mean a thin coating of any substance, such as a veneer of varnish.

Associations

A **caul** is a tool which is used to shape veneer to a curved surface. A **veneer saw** is a small back saw with fine teeth which is used for cutting veneers and finishing veneer work.

See the furniture of the great English 18th century cabinet-makers Chippendale (1705 - 1779), Hepplewhite (? - 1786) and Adam (1728 - 1792), all of whom used veneers in their work.

See: **accelerator, adhesive, face, figure, hammer, hardboard, key, laminate, lipping, marquetry, particleboard, plywood, saw.**

Viscosity

Pronounced: VIS-KOS-ITY (*i's as in his, o as in got, y as in duty*)

Origin

From the Latin *viscosus* meaning *full of bird lime and sticky.*

Meaning

Viscosity means the extent to which a thick, sticky liquid will flow when poured. Liquids of high viscosity will pour and move slowly; those of low viscosity will pour and move easily. Generally, the larger the molecule of the substance, the greater is the viscosity.

The viscosity of a liquid is sometimes referred to as its **consistency**.

The adjective from viscosity is **viscous**. A **reducer** is a volatile substance used to decrease the viscosity of a finishing medium.

Associations

See: **lacquer, paint, resin, varnish.**

Vulcanise

Pronounced: VUL-KAN-IZ (*u as in bonus, a as in ran, i as in fine*)

Origin

Vulcan was the god of fire in Roman mythology. The word vulcanise was coined by William Brockedon, a friend of the English chemist Thomas Hancock, after the vulcanising process had been discovered (by chance) by an American, Charles Goodyear, in 1838.

Meaning

Vulcanisation is a process to produce a **semi-synthetic** material. Latex or natural rubber is *heated* (See: *vulcan* above) with sulphur or sulphur compounds, which causes crosslinking of the rubber molecules. This results in the rubber becoming hard but resilient and able to retain its shape and have some elasticity. It can be moulded. Previously, rubber had been soft and sticky when heated but brick-hard when cooled. Tyres are made from vulcanised rubber.

Associations

See: **adhesive, cross-linking, moulding, shape, synthetic.**

Warp

Pronounced: WORP (*o as in port*)

Origin

From the Old English *weorpan* meaning *to throw* and also *to twist violently out of shape*.

Meaning

Warp refers to the twisting and distortion of wood ,such as the spiral twisting of the ends of wood boards or the buckling of wooden picture frames. To prevent the warping of picture frames, museums and Art galleries maintain an atmosphere in their buildings which has about 55% humidity.

Warping also refers to the distortion of a plastic moulding owing to the removal of the plastic from the mould before the plastic is set or cured.

In foundry work, warping describes the twisting of a casting, owing to unequal strains in different parts of the casting during the cooling process.

Associations

See: **casting, cleat, curing, distortion, fabric, hardening, key, kiln, moulding, season, shrinkage.**

Web

Pronounced: WEB (*e as in met*)

Origin

From the Old English *web*, which originated with an earlier Teutonic word meaning *to weave*. To web came to mean to spread out in a fine, thin form, as in a spider's web.

Meaning

Webbing is said to occur when plastic materials are not drawn satisfactorily into a mould, owing to a poorly-designed mould, and there is distortion in the moulded objects produced.

Associations

See: **distortion, moulding.**

webbing for the seat of a chair

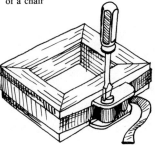

Webbing cramp

Weld

Pronounced: WELD (*e as in fell*)

Origin

From the Old English *welled* from the verb *well* meaning *to spring, boil up, rise* and, following the idea of a liquid rising, *to bring metal to a fluid state*.

Hammer or forge welding was practised in Middle East countries before 1350 B.C. and it remained the form of welding until the late 1800s. In 1877, Elihu Thomson, an American engineer, introduced electrical-resistance welding, which is the foundation of modern welding processes. Oxyacetylene welding was developed in France in the early 1900s.

Gas welding.

Gas metal arc welding

Manual metal arc welding.

Kind permission of The Welding Institute, Abington, UK.

Meaning

Welding is one of the joining processes. The others are *riveting*, *bolting*, *glueing* and *adhesive bonding*.

There are more than sixty welding processes in three broad groups, namely, **fusion welding**, **solid-state bonding** and **brazing and soldering**. However, some metallurgists do not consider brazing and soldering as welding. To weld is the process of heating two separate pieces of metal until their edges are molten and then fusing the molten sections together. This is **fusion welding** and the process is called **coalescence**. **Forge welding** is when metal is heated to a molten state, then the parts to be joined are placed one on top of each other and then hammered until they fuse. **Gas welding** (also called **oxyacetylene welding**) uses an intense flame (approximately 3000°c) from a mixture of burning oxygen and acetylene gas (composed of two parts of carbon and two parts of hydrogen) to heat **welding rods** which are placed where two pieces of metal are to be joined. The rods melt and fuse into the two pieces of metal, thus joining them. **Arc welding** is a process where a high-amperage electric current (both direct current and alternating current may be used) is passed between a metal to be welded and an electrode or rod. Electricity *arcs* between the edges of the metals to be joined, generating sufficient heat for the metals to melt. Additional metal is also melted into the joint and the metals are bonded. **Spot welding** is a process where metals are clamped between two electrodes and an electric current is passed between them, which generates sufficient heat at a contact spot to join the metals.

In recent years, **electronic-beam welding** and **laser-beam welding** have been developed, and these methods extend the precision and speed of welding processes. Robot welders, controlled by automation, are now widely used in industry, especially in motor car manufacture.

Hot air welding is used to join pieces of thermosetting plastics.

Associations

Acetylene is a volatile hydrocarbon gas, which is colourless, and has an unpleasant smell. It is used as a fuel for welding, soldering and cutting metals. It produces one of the highest flame temperatures possible. To ensure acetylene is safe during storage in a cylinder, it

is dissolved in **acetone**. Acetylene and oxygen are stored in separate steel cylinders, the oxygen cylinder being black and the acetylene cylinder being maroon. A **manifold** is the connection of a number of welding gas-cylinders by a common supply line. **Welding flux** is a compound used in welding to clean metals and prevent oxidation, to assist metal flow and to assist fusion of metals.

See: **bead, bevel, brazing, electrode, filler, fillet, fusibility, flux, jig, joint, lap, laser, plastic, sculpture, solder, thermoplastic, thermosetting**.

Whetting

Pronounced: WET-ING (*e as in met*)

Origin

From Old English *hwettan* meaning *to sharpen*. The use of stones as sharpening instruments goes back to the times when people lived in caves.

Meaning

Whetting is the process of sharpening an instrument or tool by rubbing it on or with a stone. A **whetstone** is a stone which is made of abrasive materials and then shaped so that it can be used as a sharpening device. An **Indian Oilstone**, for instance, is a whetting stone made from aluminium oxide abrasive which is impregnated with oil. They are available in three degrees of coarseness: fine, which is mainly for polishing and medium and coarse which are mainly for cutting.

Associations

See: **abrasive, shape.**

Wrench

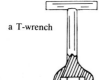

a T-wrench

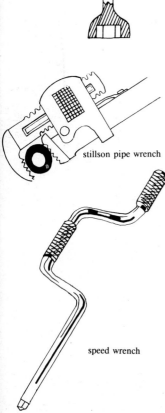

stillson pipe wrench

speed wrench

Pronounced: RENSH (*e as in red*)

Origin

From the Old English *wrencan* meaning *to twist*.

Meaning

A wrench is a tool, which is used to grip and twist and turn nuts and bolts.

Open-end wrenches are available in sets in the metric system from 7 to 19mm. **Box wrenches**, where the box is usually a double hexagon, has twelve grooves which engage the corner of a nut, so that the wrench can move through an arc of 30° before it needs to be repositioned. A **Socket wrench** (for example a **T-wrench**) which like a box wrench surrounds a nut, can prevent the tool from slipping off a nut The **ratchet wrench** allows a wrench to move through a small arc before returning to its original position, which is useful where it is impossible to completely rotate a wrench. Some wrenches are adjustable (e.g. a **monkey wrench** and a **stillson wrench**, which is usually used on metal pipes), so that they can be used to tighten or loosen nuts and bolts of various sizes. A **torque wrench** has a gauge which shows how much a nut is being *torqued* or tightened. It is used where a specified tightness is needed, as, for instance, on cylinder head bolts of a car engine.

Associations

See: **alloy, bolt, nut, torque**